Stereochemistry

Nobel Prize Topics in Chemistry
A Series of Historical Monographs on Fundamentals of Chemistry

Editor
Johannes W. van Spronsen
(The Hague and University of Utrecht)

Advisory Board
G. Dijkstra (Utrecht)   N. A. Figurowsky (Moscow)
F. Greenaway (London)   A. J. Ihde (Madison)
E. Rancke-Madsen (Copenhagen)   M. Sadoun-Goupil (Paris)
Irene Strube (Leipzig)   F. Szabadvary (Budapest)
T.J. Trenn (Munich)

Nobel Prize Topics in Chemistry traces the scientific development of each subject for which a Nobel Prize was awarded in the light of the historical, social and political background surrounding its reception. In every volume one of the Laureate's most significant publications is reproduced and discussed in the context of his life and work, and the history of science in general.

This major series captures the intellectual fascination of a field that is too often considered the domain of specialists, but which nevertheless remains a significant area of study for all those interested in the evolution of chemistry.

Current titles
Stereochemistry   *O. Bertrand Ramsay*
Transmutation: natural and artificial   *Thaddeus J. Trenn*
Inorganic Coordination Compounds   *George B. Kauffman*

Stereochemistry

O. Bertrand Ramsay

Eastern Michigan University, Ypsilanti, Michigan, USA

LONDON · PHILADELPHIA · RHEINE

Heyden & Son Ltd, Spectrum House, Hillview Gardens, London NW4 2JQ, UK
Heyden & Son Inc., 247 South 41st Street, Philadelphia, PA 19104, USA
Heyden & Son GmbH, Devesburgstrasse 6, 4440 Rheine, West Germany

British Library Cataloguing in Publication Data

Ramsay, O B
  Stereochemistry. — (Nobel prize topics in chemistry).
  1. Stereochemistry
  I. Title II. Series
  541'.223 QD481

  ISBN 0-85501-681-7
  ISBN 0-85501-682-5 Pbk

Printed in Great Britain by Cambridge University Press, Cambridge

Contents

List of Plates   vii
Foreword   ix
General Editor's Preface   xi
Author's Preface   xiii
Acknowledgements   xv

1. Key Papers by Odd Hassel, Derek H. R. Barton, Vladimir Prelog
    and John W. Cornforth   1
        Biographical notes and commentary on the significance of
        the articles   1
        O. Hassel, 'The Cyclohexane Problem'   10
        Derek H. R. Barton, 'The Conformation of the Steroid
        Nucleus'   16
        Vladimir Prelog, 'Newer Developments of the Chemistry of
        Many-membered Ring Compounds'   25
        John W. Cornforth, 'Asymmetric Methyl Groups, and the
        Mechanism of Malate Synthase'   38

2. Early Geometrical Concepts of Matter   43

3. Valency and Chemical Structure   52

4. Optical Activity and Stereoisomerism   69

5. The Origin of Stereochemistry: the Contributions of
    J. H. van't Hoff and J. A. Le Bel   81

6. The Stereochemistry of Addition and Elimination Reactions   98

7. The Stereochemistry of Substitution Reactions   107

8. Symmetry, Asymmetry, Chirality   116

9. The Specification of Molecular Configuration   122

10. The Search for Chiral Centers other than Carbon   129
    A. Trivalent nitrogen   129
    B. 'Pentavalent' nitrogen   133
    C. Other heteroatoms   140

11. The Stereochemistry of Trivalent Carbon Species   142
    A. Carbonium ions (carbocations)   143
    B. Carbon 'free radicals'   145
    C. Carbanions   146

12. Rotational 'Barriers' About Single Bonds and Steric Effects   148
    A. Rotational barriers   148
    B. Steric hindrance   150
    C. Atropoisomeric compounds   153

13. The Stereochemistry of Cyclic Compounds: the Early History   159

14. The Origins and Development of Conformational Analysis   177
    A. Acyclic compounds   177
    B. Cyclic compounds   183

15. Asymmetric Transformations   194

16. Some Recent Developments and Future Expectations   202
    A. 'Chemical curiosities'   202
    B. Biological stereochemistry   214

Appendix A.   Chronology of Events and Publications in the History
              of Stereochemistry   226

Appendix B.   Glossary   231

Appendix C.   Stereochemical Satire   236

Appendix D.   Supplementary Reading   245

Subject Index   249
Name Index   253

List of plates

		Page
Plate 1.	Odd Hassel	2
Plate 2.	Derek H. R. Barton	4
Plate 3.	Vladimir Prelog	6
Plate 4.	John W. Cornforth	7
Plate 5.	Friedrich August Kekulé von Stradonitz (1829–1896)	54
Plate 6.	Jean Baptiste Biot (1774–1862) (Edgar Fahs Smith Collection)	70
Plate 7.	Louis Pasteur (1822–1895) in 1857 (Institut Pasteur)	72
Plate 8.	Johannes Adolf Wislicenus (1835–1902)	79
Plate 9.	Jacobus Henricus van't Hoff (1852–1911) (Edgar Fahs Smith Collection)	82
Plate 10.	Joseph Achille Le Bel (1847–1930) (Edgar Fahs Smith Collection)	88
Plate 11.	Hermann Kolbe (1818–1884)	92
Plate 12.	Paul Walden (1863–1958) (Edgar Fahs Smith Collection)	107
Plate 13.	Johann Friedrich Wilhelm Adolf Baeyer (1835–1917) (Edgar Fahs Smith Collection)	159

Foreword

By D.H.R. Barton, FRS
(Nobel Laureate in Chemistry, 1969)

In the daily struggle to advance the frontiers of Science, research workers often neglect the history of their speciality. This is regrettable because history, and especially the history of ideas, has much to teach us about the psychology of creative thought.

Every organic chemist recognizes the fundamental importance of structural theory. This is expressed in the concept of constitution, configuration and conformation. It is a fascinating exercise to read how each of these ideas came to be proposed, discussed and finally accepted. In each case the concept was needed to correlate an expanding body of experimental fact. In each case there was surely a large number of chemists who could have proposed the concept but only one or two who did.

The particular case of conformational analysis is now recent history. Here one has to ask why conformational analysis was not recognized much earlier. The work of Hermans and of Böeseken on ring formation from diols or the work of Reeves on carbohydrates is certainly early conformational analysis. But in each case the general chemical public could not appreciate the implications of the papers. Perhaps the time was not ripe. When conformational analysis was finally generally accepted, it was because it was much needed in steroid chemistry where the cortisone problem created hundreds of would-be steroid chemists almost overnight. At the same time the physical methods of electron diffraction, of X-ray crystallography and of calorimetric and computation entropy analysis had reached the point where organic chemists could believe that the conclusions reached were true. In case the reader should think that everything was obvious, it must be pointed out that in the literature for

1935–1950 there are two authoritative papers by well-respected scientists which show that the eclipsed conformation of ethane is *more* stable than the staggered and that the boat conformation of cyclohexane is *more* stable than the chair. Even today the writings of very distinguished chemical physicists continue to confuse configurations with conformations!

This book by Dr Ramsay is a clear, well-written and very interesting account of the history of organic chemical ideas on stereochemistry. It is to be recommended not only to students of the subject but also to active research workers in organic chemistry. Both groups have much to learn about creative thinking from the analysis of the past.

Paul Barton

CNRS,
Institut de Chimie des Substances Naturelles,
France

General Editor's Preface

Nobel Prizes in Chemistry have been awarded almost every year since 1901, and the topics covered by these awards have touched upon almost every subject in chemistry. 'Nobel Prize Topics in Chemistry' plans to cover the history of each subject for which a Nobel Prize was awarded and to place particular emphasis on the life and work of the Nobel Prize winner himself. In this way the planned Series will come to describe the whole history of chemistry. The concept is to take one of each Nobel Laureate's most significant publications, to reprint it (as an English translation if appropriate), discuss it, and then to place it within the context of the Laureate's life and works in particular and the history of science in general, if possible going as far back as Egyptian, Babylonian and Greek antiquity. The Series will also look at possible future developments.

Contributions are presented in such a manner that a non-specialist background will suffice for the text to be comprehensible. The intention is to make as many readers as possible aware of and conversant with the problems underlying the development of various areas in the field of chemistry. Each volume attempts to give a smooth outline of the particular topic under consideration, uninterrupted by continual footnotes and references in the text.

The Series is not, in the first instance, aimed at the professional historian, but rather at the chemist, the research worker and the non-specialist who wishes to bring himself up to date on the historical background of one or more areas of chemistry. The student of chemistry and the historian or sociologist who for research wishes to focus broadly on one of the most spectacular disciplines in the natural sciences, can obtain a wide-ranging historical knowledge – knowledge which forms part of the general history of mankind and which can also be used to examine the reciprocal relationship between chemistry and society as a whole. The authors are well-known historians of chemistry or chemists with a solid knowledge of history. Among the latter group, occasionally a Nobel Prize winner will be our author. Nobel Prize winners are also contributing forewords to many of our volumes.

For certain parts of the texts a reasonable knowledge of chemistry – in some cases, the reading of formulas for example – is required, but the needs of the non-chemist have been anticipated by including in each volume a glossary through which the reader can revise or extend his chemical knowledge. Each volume also contains a chronology of significant events and a detailed bibliography.

It is the aim of the Editor that the readers of this Series should obtain a clear idea of the particular experiences of a chemist in performing his research – research which sometimes led to a discovery of the greatest significance for humanity. It is thus not only the intention to focus exclusively here on the main historical-chemical facts, but also to under-stand the chemist as a human being and look at the circumstances which led to his discoveries. However, to understand the social and eventually the political background of these historical developments, the reader must inform himself of the facts presented here. This Series thus aims to capture the intellectual fascination of a field that is too often considered to be the domain of specialists, but which nevertheless remains an area of proven intellectual adventure for all those who consider the quest for understanding the highest point to which man can aspire.

It is our sincere wish that the individual volumes in this Series shall realize these aims and intentions.

J. W. van SPRONSEN
The Hague

Author's Preface

The award of Nobel Prizes in Chemistry in 1969 made jointly to Odd Hassel and Derek H. R. Barton, and in 1975 to John W. Cornforth and Vladimir Prelog, mark the recognition of major turning-points in the history of stereochemistry. In order to understand the importance of the research that led to these awards, a major portion of this book will be concerned with the development of stereochemistry from the middle of the 19th century until the 1950s when stereochemical research underwent a significant change in emphasis. Later chapters examine the nature of this change and attempt a look into the future of stereochemical research. The first chapter contains biographical and other information about the Nobel Laureates in order that the reader can more fully appreciate the significance of the four key papers that follow. The reader with only a marginal background in stereochemistry may wish to defer the reading of these papers until the completion of the remaining chapters.

The early chapters discuss some of the theoretical and experimental studies undertaken in the early part of the 19th century which finally led to the simultaneous proposals of J. H. van't Hoff and J. A. Le Bel in 1874, a date that might be considered to mark the birth of stereochemistry. The designation of birthdates is perhaps historically irrelevant, but it would seem that 1874 does mark the beginning of a paradigmatic shift in chemistry equal in significance to the introduction of the theory of valency by August Kekulé and A. S. Couper in 1858. The changes brought about by the publications of Hassel, Prelog and Barton in the early 1950s, while dramatic, cannot be considered paradigmatic. The contrast can be seen in the reception of the ideas. The proposals of Barton, in particular, were almost immediately adopted and exploited by a large number of chemists; the proposals of Van't Hoff and Le Bel, on the other hand, were initially greeted with indifference and sometimes with hostility. It would seem that in this case there were not a sufficient number of prominent chemists actively interested in the problems to which the proposals of

Van't Hoff and Le Bel were directed. Throughout the book I have concentrated on the internal development of stereochemistry, and only considered in a superficial way the impact of stereochemistry on research in other areas such as biochemistry and inorganic chemistry, an exception being made as regards the inclusion of a discussion of studies concerned with the stereochemistry of biosynthetic reactions.

The organization of the topics in this book might strike the historian of science as somewhat unconventional. The reason for this is that a secondary aim of the book is to introduce the reader to some of the basic concepts and terminology of stereochemistry. Although it is assumed that the reader will not have an extensive background in either chemistry or the history of chemistry, it is to be hoped that the specialists in these areas will not find the book without interest.

The first draft of this book was written while on Sabbatical leave from Eastern Michigan University in 1973–74. The support of the University during this leave is gratefully acknowledged. Library and study facilities were generously provided during my stay at the University of Reading, England. I should like to thank Professors Andrew Gilbert, Neil Isaacs, Ernest Halberstadt, Derek Bryce-Smith, George Esselmont and Jim Irwin for their encouragement and for making my first trip to England such a pleasant one.

I am particularly grateful for the encouragement given to me by Professor Ernest L. Eliel, who devoted considerable time to a critical reading of the first draft. I should also like to express my appreciation to my wife, Patricia, for her continued support and encouragement.

June 1981

O. BERTRAND RAMSAY
Department of Chemistry
Eastern Michigan University
Ypsilanti, Michigan 48197
USA

Acknowledgements

Thanks are due to the following for permission to reproduce copyright material: Professor Odd Hassel, Dr Kenneth Hedberg and John Wiley & Sons, Inc., for the English translation of 'The cyclohexane problem', in N.L. Allinger and E.L. Eliel (eds), *Topics in Stereochemistry,* Vol. 6, pp. 11–12 (Wiley-Interscience, New York, 1968); Sir Derek Barton, F.R.S., and Birkhäuser Verlag (Basle, Switzerland) for 'The conformation of the steroid nucleus', *Experientia* **6**, 316–320 (1950); Professor Vladimir Prelog and the Royal Chemical Society for 'Newer developments of the chemistry of many-membered ring compounds', *Journal of the Chemical Society* 420–428 (1950); Sir John Cornforth, C.B.E., F.R.S. and Macmillan Journals Ltd for 'Asymmetric methyl groups, and the mechanism of malate synthase', *Nature (London)* **221**, 12–13 (1969); Mr William Graham and the Society of Chemical Industry (London) for the Letter to the Editor, from *Chemistry and Industry* p. 1533 (25 August 1962); Professor John T. Edward and the Society of Chemical Industry (London) for Fig. 130, from *Chemistry and Industry* p. 354 (26 March 1955); Professor Robert E. Lyle and the Division of Chemical Education, American Chemical Society, for Fig. 131, from *Journal of Chemical Education* **50**, 655–656 (1973); Punch Publications Ltd for Fig. 132, from *Punch* p. 69 (16 September 1959).

To

Patricia and Sean

1

Key papers by Odd Hassel, Derek H. R. Barton, Vladimir Prelog and John W. Cornforth

BIOGRAPHICAL NOTES AND COMMENTARY ON THE SIGNIFICANCE OF THE ARTICLES

In 1969 the Nobel Prize in Chemistry was awarded jointly to Odd Hassel, of the University of Oslo, and Derek H. R. Barton, of the Imperial College of Science and Technology (London), 'for developing and applying the principles of conformation in chemistry'. The term conformation might be defined as the three-dimensional arrangements of a molecule that are possible by virtue of rotation about single bonds.

Conformation ideas form such an integral part of present-day discussions of organic chemistry that it is perhaps difficult to appreciate the importance and impact of the introduction of such ideas in the middle of the 20th century. A dramatic shift in the direction of stereochemical research took place as a result of the publication of Barton's paper in the journal *Experientia* in 1950. Barton was able to apply the results of earlier physical chemical studies concerning the conformation of molecules to show how the conformation of a molecule determined some of its physical and chemical properties. The introduction of conformational analysis, which is the study of this interrelationship, into chemical research has been considered by many chemists as representing the first real advance in stereochemistry since the introduction of the theory of Van't Hoff and Le Bel in 1874. Prior to 1950 stereochemical research was largely concerned with the investigation of the phenomena of stereoisomerism. The change that took place in stereochemical research may be illustrated with reference to the history of the structure of one particular compound, cyclohexane. The historical details of the development of the studies concerned with the structure and conformation of cyclic compounds are provided in Chapters 13 and 14.

Prior to 1890, cyclohexane was generally considered as having a planar hexagonal arrangement of the carbon atoms. This view derived in large part from the influence of the Baeyer 'strain' theory. In 1890 H. Sachse challenged this view by showing how, with tetrahedral carbon models, two strain-free forms of cyclohexane could exist. One was a somewhat rigid, chair-shaped form; the other a flexible, boat shape. Since the experimental studies undertaken by other chemists to test the validity of Sachse's hypothesis were inconclusive, his ideas remained dormant for some 30 years. In 1918 E. Mohr revived the hypothesis by showing how the existence of numerous bi- and tricyclic hydrocarbons could not be accounted for in terms of planar rings. In his paper he suggested that the existence of a multiplanar cyclohexane ring might be proved by the isolation of isomers of the bicyclic hydrocarbon, decalin. Although W. Hückel was able to synthesize two decalin isomers seven years later, many chemists were still unwilling to concede that the simpler monocyclic rings need be multiplanar. Certainly the chemical methods employed in the 1920s and 1930s to decide this question produced ambiguous results. Most organic chemists continued to illustrate cyclohexane with a planar structure since such a structure could satisfactorily account for the number of known stereoisomers. More persuasive evidence for the multiplanar ring emerged as the result of physical chemical investigations. The studies of Odd Hassel in Norway played a key role in the shift of chemists' views on the structure of cyclohexane.

Odd Hassel (Plate 1) was born in Christiania (now Oslo), Norway, on 7 May 1897. He graduated from the University there in 1920, having studied mathematics and physics, with chemistry being his main subject.

Plate 1. Odd Hassel.

He received his doctorate of philosophy degree from the University of Berlin in 1924. It was in this period that he was introduced to the techniques of X-ray spectroscopy. He joined the faculty of the University of Oslo in 1925 and remained there ever since. He became the first head of the department of physical chemistry in 1934. Beginning in about 1930, Hassel used dipole moment measurement and X-ray diffraction spectroscopy to study the structure of cyclohexane and its derivatives. These methods were of limited applicability, however. Dipole moment measurements do not lead to a unique structural determination. As is discussed in Chapter 14, for example, the observation of a dipole moment may rule out the existence of a molecule in one particular conformation, but it cannot decide on what conformations are actually present and in what equilibrium concentrations. Although X-ray measurements provided more structural information, they were confined to compounds that could be obtained in the solid state. In 1938 Hassel was able to obtain much more complete structural information from electron diffraction measurements of cyclohexane and cyclohexane derivatives in the gas phase. Hassel's earlier work appeared in German journals, but with the outbreak of the war he began to publish in a relatively obscure Norwegian journal. Thus his most important article published in 1943, and which is reprinted in translation here, was unknown to most physical chemists until much later. The 1943 paper summarized the evidence that he had obtained to show that cyclohexane existed largely in the chair form. In monosubstituted cyclohexane derivatives the substituent could be found in one of two positions (the equatorial or axial) which could be interconverted by a rapid ring inversion, although the conformation in which the substituent was in the equatorial position was the more stable.

Hassel was arrested by the Norwegian Nazis shortly after the publication of this paper and on his release in 1944 found the laboratories almost completely abandoned. After the war he continued his earlier studies on the conformations of acyclic, cyclic and bicyclic compounds. His work became better known in 1946–47 as the result of the publication of several English-language articles in the journals *Nature* and *Acta Chemica Scandinavica.* He died in May 1981 aged 83.

Derek Barton (Plate 2) was one of the chemists who read these articles and saw their relevance to a number of stereochemical problems in organic chemistry. Derek H. R. Barton was born on 8 September 1918 in Gravesend, England. After attending Gillingham Technical College, he proceeded to Imperial College from which he received his B.Sc. degree in 1940 and Ph.D. in 1942. Afterwards he was involved for three years in war research, which led him to the development of a catalytic method of preparing vinyl chloride from ethylene dichloride. It was during this period that he used his spare time to develop a theory of molecular rotation of triterpenoids and steroids. As a result he became quite familiar with the

Plate 2. Derek H.R. Barton.

chemistry of natural products. After the war he returned as an Assistant Lecturer to Imperial College to lecture in inorganic chemistry and physical chemistry. Because of his reading in the latter field he became aware of Hassel's papers.

In 1949 Barton was invited to Harvard University as a visiting lecturer to replace R. B. Woodward (Nobel Prize in Chemistry, 1965) who was on a Sabbatical leave. It was during this period that he wrote the paper which is reprinted here. Barton recounted that

the paper on conformational analysis was written because I had a good knowledge of steroid chemistry and triterpenoid chemistry, and a good knowledge of organic chemistry, and at the same time, since I had been doing physical chemistry and teaching it, I had been reading the literature of physical chemistry and of chemical physics. Few organic chemists in those days were able or inclined to read both kinds of literature. The paper takes what had been done by chemical physicists, particularly by Hassel, and applies their discoveries to an analysis of the organic chemistry of steroids and polyterpenoids – complex natural products which the chemical physicist would never have thought of looking at.

The paper was written after he had attended a seminar given by Louis Fieser at Harvard. Fieser, a well-known steroid chemist, had been describing his difficulty in understanding certain relationships between the relative ease of hydrolysis of esters in several steroids (see Chapter 14). Barton says: 'I knew at once, from having the shape of the molecules in my mind, how this phenomenon could be explained.' The paper demonstrates how a knowledge of the conformation of such complex molecules could be used to explain the stereochemistry of a number of reactions or deduce the configuration of a compound. The publication of this paper had a profound and immediate impact on the subsequent development of natural products chemistry. Soon the concepts of conformational analysis were incorporated into nearly all branches of organic chemistry and biochemistry. Some of these developments are discussed in more detail in Chapter 14. In a second paper published in 1953, Barton expanded the applications of conformational concepts to the explanation of physical properties such as absorption affinity and infrared spectra. It is difficult to conceive what might have been the subsequent development of organic chemistry and biological chemistry in the absence of Barton's paper. Even though both Hassel and Barton ceased active research in the field of conformational analysis within the next decade, the development and application of the concepts first proposed has continued at a rapid pace.

The 1975 Nobel Prize in Chemistry was awarded jointly to Vladimir Prelog and John Cornforth for their researches on the stereochemistry of organic molecules and reactions, and the stereochemistry of enzyme-catalyzed reactions, respectively.

Vladimir Prelog (Plate 3) was born in Sarajevo, Yugoslavia, on 23 July 1906. He received his chemical training at the Technical University in Prague, Czechoslovakia, receiving his doctorate in 1929. Prior to 1941 he was associated with the G. J. Driza Laboratory in Prague and the Chemistry Department at the University of Zagreb. Since 1942 he has been on the staff of the Laboratory of Organic Chemistry at the Eidgenössische Technische Hochschule in Zürich, Switzerland.

Much of Prelog's research was initially concerned with the chemistry and stereochemistry of natural products. Of particular interest to the readers of this book might be his synthesis of the hydrocarbon adamantane and Troeger's base (discussed in Chapter 10). Prelog's synthetic and structural studies led him to demonstrate the limitations of 'Bredt's Rule' concerning the location of double bonds at bridgehead carbons in bicyclic systems (Chapter 13).

When Prelog arrived at Zürich in the early 1940s, he first undertook the synthesis of medium-sized (8- to 11-carbon atom rings) carbocyclic rings. Leopold Ružička had recently received the Nobel Prize in Chemistry (1939) for his synthesis of the macrocyclic compounds muscone and

Plate 3. Vladimir Prelog.

civetone. By 1947 Prelog and a colleague, M. Stoll, had synthesized a number of the medium-sized rings and had initiated several studies to ascertain the relationship between ring size and reactivity. These studies were summarized in a lecture given to the Chemical Society in London on 17 February 1949. The lecture was published as an article in the *Journal of the Chemical Society* in January 1950. It was in this paper, reproduced here, that Prelog discussed the relationship between a ring's 'constellation' (now: conformation) and its reactivity. For example, it was observed that whereas cyclohexanone ($n = 6$ below) forms a cyanohydrin, the cyanohydrin of cyclodecanone ($n = 10$) is not formed at all (see Figure 5 in the reprinted paper).

$$[CH_2]_{n-1} \quad HO-C-CN \quad \rightleftharpoons \quad [CH_2]_{n-1} \quad C=O \quad + \quad HCN$$

Cycloalkanone cyanohydrin   Cycloalkanone

The ease of formation of the cyanohydrin could be correlated with a particular 'constellation' of the ring. The term 'constellation' was defined by Prelog as describing '. . . those forms of the molecules which result from free rotation around single bonds, for example, the chair and the boat form of cyclohexane'. The relationship between the molecular conformation (constellation) and chemical reactivity is discussed in some detail and set the ground work for further studies in this field. Unfortunately, Prelog's assumption of a special kind of intramolecular hydrogen bond-

Plate 4. John W. Cornforth.

ing (see Figures 9 and 10, for example) was not supported by subsequent studies. The paper also contains a summary of the application of Bredt's Rule to bicyclic systems.

Although Prelog's paper was published several months earlier than Barton's, it did not receive the attention afforded the latter's, presumably because chemists concerned with natural products seldom encountered cyclic compounds containing more than 6 or 7 carbon atoms.

The paper by John W. Cornforth (Plate 4) reprinted here illustrates a somewhat different aspect in the development of stereochemistry. John W. Cornforth was born in Sydney, Australia, on 7 September 1917. He received his doctoral degree in 1941 at Oxford University in England working under Sir Robert Robinson (Nobel Prize in Chemistry, 1947). Cornforth's collaboration with Robinson continued until 1946 at which time he moved to the Mill Hill Research Laboratories of Britain's Medical Research Council. It was in this period that he initiated his research on the stereochemistry of enzyme processes – in particular, the biosynthesis of squalene from mevalonic acid. It was for this work that he was recognized in the award of the Nobel Prize in 1975.

The details of the squalene biosynthesis are discussed in Chapter 16. The present discussion will serve as an introduction to a paper published in 1969 and which is reprinted here. The synthesis of squalene from mevalonic acid involves 14 stereochemically controlled reactions. By 1969, the stereochemistry of all but one of the steps had been determined by Cornforth and his co-worker, G. Popjak. The question that remained related to the isomerization of isopentenyl pyrophosphate to 3,3-dimethylallyl pyrophosphate. It was not clear as to the stereochemistry involved in the addition of the hydrogen to carbon number 4 — that is, did it add from above the plane of the paper (in the reaction below) or above?

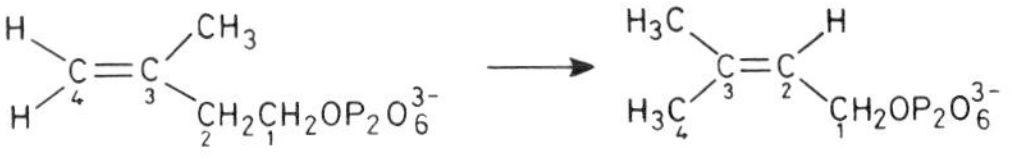

Isopentenyl pyrophosphate   3,3-Dimethylallyl pyrophosphate

This question may seem trivial indeed since the stereochemistry of the product will not provide the answer to the question. This reaction illustrates the 'hidden stereochemistry' involved in many enzymatic reactions — that is, there is stereochemical control even though it is not revealed by an examination of the reactants and products. Several illustrations of this hidden stereochemistry are discussed in Chapter 16. For example, the enzymatic oxidation of ethanol to ethanal involves the removal of only one of the two available hydrogens on carbon number 1:

Ethanol   Ethanal

The only way the stereochemistry of the isomerization reaction could be determined required the use of isotopic hydrogen. In 1972, Cornforth had determined that the stereochemistry proceeded as illustrated below when tritium was added to 4-deuteroisopentenyl pyrophosphate:

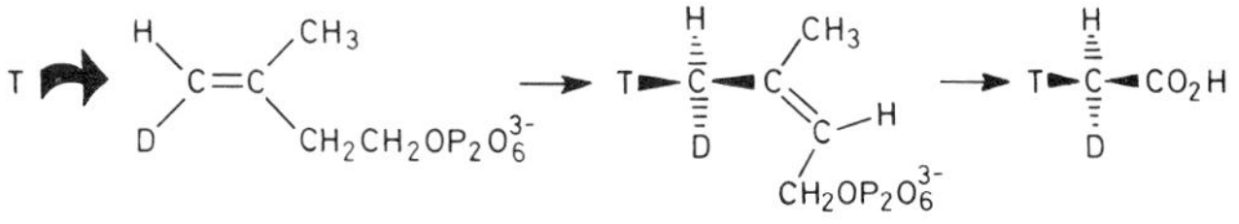

4–Deuteroisopentenyl pyrophosphate   (S)–Deuterotritioacetate

The stereochemistry of the product was established by converting the product to the chiral form of acetic acid. However, the 1972 paper required a knowledge of the absolute configuration of the enantiomeric deuterotritioacetates. The 1969 paper reprinted here provided an elegant means by which this could be accomplished. Briefly summarized, Cornforth devised a stereospecific method of synthesizing the acetates of a known configuration (Procedure III to IXa and IXb in the article). The paper also provided another example of hidden stereochemistry. In the preparation of malic acid from acetate (as acetyl coenzyme-A), it was found that the displacement of a hydrogen from the methyl group in the acetate proceeded by an inversion process. This inversion is illustrated in the last figure in the paper. (At ETH in Zürich, D. Arigoni and J. Rétey independently produced a similar solution to the same problem.)

The work of Cornforth signaled the beginning of the realization of the importance of stereochemical concepts to the understanding of biological systems.

THE CYCLOHEXANE PROBLEM

O. HASSEL

[translated from *Tidsskrift for Kjemi, Bergvesen og Metallurgi* 3 (5), 32 (1943), by Kenneth Hedberg, Oregon State University, Corvallis, Oregon; from N. L. Allinger and E. L. Eliel (eds.), *Topics in Stereochemistry*, Vol. 6, pp. 11–17. Wiley-Interscience, New York (1971)]

If one assumes that the valence angles of the carbon atom are equal to the 'tetrahedral angle' ($109°28'$), or in any case not substantially different from this angle, then the possibility that the carbon atoms in cyclohexane form a coplanar six-membered ring is excluded. Among the forms of the carbon skeleton compatible with an angle in the vicinity of $110°$ the 'chair' form occupied a special position. In this form the cyclohexane molecule has ditrigonal skalenohedral symmetry (symbol D_{3d}) in which it has, besides a threefold symmetry axis, a center of symmetry (Fig. 1a). This form is 'stiff' in the sense that a transformation into other forms is impossible without significant deformations of the valence angles. According to the knowledge we have at present about the forces between carbon atoms in compounds of this type, it seems very likely that the molecule must be provided with a not inconsiderable amount of energy in order to be converted into the other conceivable strain-free forms. On the other hand, it is hardly possible, *a priori*, to say anything definitive about the energy difference between the cyclohexane molecule in the symmetric chair form and in these other forms, even though it might seem improbable that this energy difference is considerable. One has been inclined to the opinion, therefore, that cyclohexane in thermodynamic equilibrium will contain perceptible amounts of molecules with lower symmetry than the chair form.

In order to clarify this point one has had to turn to physical-chemical methods of investigation and, in addition to studying the hydrocarbon itself, it was natural to take up the study of a series of its simpler derivatives. Since these investigations have in large part been carried out in our laboratory, we shall, in the following, give a short characterization of the results which are currently available in the literature and report a number of new results which have not yet been published. Before we do this, however, we shall discuss briefly the views held by A. Langseth[1] concerning the cyclohexane structure. Langseth's results rest upon Raman spectroscopic investigations and were completely unexpected. They were to the effect that the molecule itself had hexagonal bipyramidal symmetry (symbol D_{6h}) and that the carbon atoms therefore formed a coplanar regular hexagon, corresponding to a C–C–C angle of $120°$. In Langseth's opinion the question of whether cyclohexane itself has symmetry D_{3d} or D_{6h} is in a way related to the question of the most stable form for the ethane molecule. If we imagine the two methyl groups in ethane rotated in relation to each other about the C–C bond, the H atoms will, in a certain position (the so-called '*cis*'[a] position), lie directly opposite each other and have the smallest possible mutual distance; by a rotation of $60°$ from this position the corresponding '*trans*'[a] form of the ethane molecule is generated where the H atoms are as distant from each other as possible. It is easy to see that the relative position of hydrogen atoms in the symmetric cyclohexane model corresponds to the ethane *trans* form, while a coplanar molecule to some degree corresponds to the *cis* form. One then disregards the circumstance that the C—C—C angle is increased from 109.5° to 120° and assumes that the energy difference between the two forms is primarily dependent upon the relative position of the H atoms. It is our opinion that such a view is not justified.[b] Nor does it find support in those results which are available concerning the orientation of methylene groups in organic compounds with long chains, where an arrangement corresponding to the *trans* position appears to be most stable. Another argument that in our opinion ought not to be overlooked is the existence of two isomeric forms of decalin.

The newer methods for electron diffraction investigation of vapors worked out in our institute allow an objective determination of the intensity curve, and by Fourier analysis one is able to calculate from this curve those internuclear distances which are most decisive for the scattering of electrons. A careful investigation of the scattering from cyclohexane[2] shows very clearly that the ratio between the carbon—carbon distances is, to say the least, very close to that which valence angles of 109°28' would give. A coplanar carbon six-membered ring is therefore ruled out. The experimental scattering curve agrees on the whole so well with the curve one calculates theoretically for molecules of the classical 'chair' form that the possibility that a significant fraction of the molecules have a less symmetric shape must be regarded as excluded. Recent Raman investigations by Saksema[3a] and by Kohlrausch and Wittek[3b] are in excellent agreement with this result. We have mentioned that A. Langseth has attempted to connect the question of the shape of the cyclohexane rings with the question of the mutual orientation of the two halves of the ethane molecule. On the basis of Raman-spectroscopic investigations of *sym*-tetrachloroethane ($CHCl_2$—$CHCl_2$)[4] he claims to have proved that the *cis* form is present in significant amounts, that is, that form where the two halves of the molecule lie symmetric to a plane perpendicular to the C—C bond, and the two pairs of chlorine atoms therefore have the smallest possible separation. This Cl—Cl distance, however, is so small that in our opinion it must result in considerable mutual repulsion, and therefore, in connection with systematic investigations of ethane derivatives, we have also studied tetrachloroethane. Through this we have been able to determine with certainty that the *cis* form, if indeed it represents a form with a minimum on the potential energy curve, must be present in very minor amounts.

In all cases where cyclohexane or its simple substitution products have been thoroughly investigated by physical methods, one has found structures containing the symmetrical valence-skeleton characteristic of the 'chair' form. As shown in Fig. 1b, six of the C—H bonds are parallel with the threefold axis of the molecule (dashed in the figure), while the six other C—H bonds form an angle of 109°28' with this direction. A bond of each type extends from each C atom. It follows from this that *two* different monosubstituted cyclohexanes with the described carbon skeleton must actually be found. Those cases where two such stereoisomeric monosubstituted derivatives supposedly have been isolated, however, scarcely bear up under closer examination. One can easily convince oneself that an inversion of the carbon ring will convert each of the isomeric compounds into the other. This conversion evidently requires such a small activation energy that a separation of the two isomeric molecules is impossible under ordinary conditions. However, it seems reasonable that the difference in energy of the two forms is, in many cases, large enough that one form will predominate in the equilibrium mixture. This is undoubtedly the case for the chloro-cyclohexanes in question, as is shown below.

If the configuration[c] of a substituted cyclohexane is to be completely described, one must specify not only the C atoms where the substitution has taken place, but also which of the two hydrogen atoms is substituted; namely, the 'erect' one, if the C—H bond is parallel with the 3-fold axis, or the 'reclining' one. In the first case we will designate the H atom and correspondingly the substituent with the letter ϵ ($\epsilon\sigma\tau\eta\kappa\omega\varsigma$) 'standing' and in the second case with the letter κ ($\kappa\epsilon\iota\mu\epsilon\nu\sigma\varsigma$) 'reclining'.[d] Some examples of halogen compounds whose configurations are determined with complete certainty are the following:

Dichlorocyclohexane	m.p. 101°	1(κ), 4(κ)
Dibromocyclohexane	m.p. 111°	1(κ), 4(κ)
Diiodocyclohexane	m.p. 142°	1(κ), 4(κ)
Tetrabromocyclohexane	m.p. 185°	1(κ), 2(κ), 4(ϵ), 5(ϵ)
β-Benzenehexachloride		1(κ), 2(κ), 3(κ), 4(κ), 5(κ), 6(κ)
β-Benzenehexabromide		1(κ), 2(κ), 3(κ), 4(κ), 5(κ), 6(κ)

In many other cases the configuration is less certain, but an unambiguous decision will often be feasible with use of modern diffraction methods. In a few cases, such as α-benzenehexachloride, the decision has proved to be difficult. An X-ray crystallo-graphic determination of the space group alone proved insufficient to answer the question, and one will be forced to carry out a complete crystal structure determination. An attempt at an electron diffraction determination carried out by O.J. Lind as his thesis problem for the *cand. real.* degree makes it seem most likely that the configuration is 1(ϵ), 2(κ), 3(κ), 4(ϵ), 5(κ), 6(κ), but 1(ϵ), 2(ϵ), 3(κ), 4(κ), 5(κ), 6(κ) and 1(ϵ), 2(κ), 3(κ), 4(κ), 5(κ), 6(κ) also seem possible. When we are concerned with halogen derivatives, space considerations exclude a number of configurations since 1,3-dihalides with both halogen atoms in the ϵ-position would bring these two atoms so near each other (2.5 Å) that such a configuration must be regarded as improbable for the dichloride and completely excluded for the dibromide and diiodide.

Of particularly great interest, naturally, is the above-mentioned question regarding the configuration of the monosubstituted derivatives. The older methods of electron-diffraction structure determination were not precise enough to reveal whether, in the

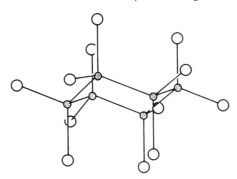

Figure 1 (a).

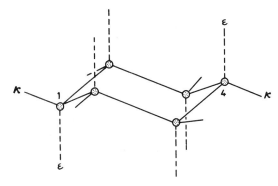

Figure 1 (b).

vapor phase, we were dealing with the κ or ϵ form of chlorocyclohexane. With the procedure now in use in our laboratory this is no longer the case. Together with *cand. real.* Henry Viervoll we have, therefore, carried out an analysis of this compound. We employed a difference method which has proved to be valuable in cases where very good photometer curves are available both for a hydrocarbon and for its monohalogen derivative. After the molecular scattering curves for both materials are determined, the difference curve is plotted and subsequently subjected to Fourier analysis. The dominant internuclear distances which are characteristic for the halogen compound but which do not appear in the hydrocarbon itself — in the case under discussion the distances between the chlorine atom and the carbon atoms of the ring — ought to appear as obvious maxima in the radial distribution curve. It turns out that besides the C—Cl distances which are common to both the κ form and the ϵ form, the curve has maxima corresponding to distances characteristic for the κ form but not for those distances which only can appear in the ϵ form. We are thus able to draw the conclusion that the amount of κ molecules is in any event considerably larger than the amount of ϵ molecules. It is thus also to be expected that the difference in energy between the two molecules is large enough to lead to a considerable excess of the one type of molecule in an equilibrium mixture. That the κ form is lower in energy and therefore the more stable form is interesting and undoubtedly will prove to be of importance for understanding the relative stability of derivatives with several halogen atoms. It appears, indeed, that substitution in the κ position takes place more easily than in the ϵ position.

In addition to the examples named above we may recall the interpretation of the configurations we have given earlier[e] for dibromocyclohexane with m.p. 45° and diiodocyclohexane with m.p. 67.5°, according to which both must be regarded to be $1(\kappa)$, $3(\kappa)$ compounds. Since the findings for these two materials are not so clearcut as the above results (p.12), we have not included them among our examples.[f] We have (also) found from an electron diffraction study by the 'visual' method that the configuration of another dibromide which is liquid at room temperature, apparently corresponds to the $1(\kappa)$, $4(\epsilon)$ compound.

The 'snapping-over' inversion of the cyclohexane ring which was mentioned in connection with chlorocyclohexane must be expected to take place in the hydrocarbon itself as well as in its derivatives. The three carbon atoms (for example 1, 3 and 5) which lie in a plane *ca.* 0.5 Å over the plane containing the three remaining carbon atoms when the molecule is oriented with the threefold axis vertical, after the process will lie 0.5 Å *lower* than these C atoms. For the hydrocarbon itself, the resulting form will be identical with the original, and one thus sees that the statistical weight of the 'chair' form must be considered doubled. For derivatives, this will ordinarily not apply, in that bonds which before the inversion were κ-bonds are converted into ϵ-bonds and vice versa. Even if the configuration before and after the mentioned trans-formation may be completely different, and in general must be expected to corre-spond to a not inconsiderable difference in energy, one easily understands that the two forms cannot be regarded as completely independent structures. According to experi-ence with chlorocyclohexane, however, one would expect that the energy difference is most often large enough that in the liquid or gas phase one form will be practically completely missing. We saw above that steric reasons will prevent a configuration with two halogen atoms in 1,3 positions both bound as ϵ atoms.

In some cases the configuration will be identical before and after the inversion, just as for the hydrocarbon itself. This applied, for example, to the above-mentioned 1,2,4,5-tetrabromide. This molecule has a 2-fold symmetry axis as its only symmetry element (symmetry C_2) and is therefore different from its mirror image. One might imagine that an inversion of the ring would in this case transform the *d*-form into the *l*-form and vice versa. This is not the case, however, as examination of a model shows. In other words, the optically active forms must be quite stable and must have a normal rate of racemization.

If one considers 1,2- or 1,4-disubstituted cyclohexanes, one must expect that the $\kappa,\kappa (= \epsilon,\epsilon)$ form corresponds to the so-called *trans* compound and the κ,ϵ form to the *cis* compound. If the 1,3-compound is considered, the question is more in doubt. We must remember that steric reasons in this case, will normally exclude the ϵ,ϵ configura-tion. The experimental data are still too sparse for us to delve into this question more deeply. We will only mention that two independent X-ray crystallographic investiga-tions of the so-called *trans* quinitol (1,4-cyclohexanediol) have given the result that the molecule must appear in a configuration characterized by a center of symmetry, some-thing which is possible only if we have a $\kappa,\kappa (\epsilon,\epsilon)$ configuration. This result therefore supports the view we have just advanced concerning 1,4-disubstituted cyclohexanes. It seems peculiar that dipole moment measurements of 1,4-cyclohexandione in solution have led to the result that this substance has an electric moment. An electron dif-fraction investigation carried out with *cand. real.* Henry Viervoll gives the result that in the gas phase the molecule has a symmetric structure very near that which one would expect for a diketone with a carbon skeleton corresponding to the chair form. The explanation of the measured dipole moment in solution must apparently be sought in a partial enolization.[g]

Since a complete summary of the older work in the literature is too lengthy to give here, we content ourselves with calling attention to references in an earlier article in this journal.[5]

LITERATURE

1. A. Langseth, *Forh. 5 Nord. Kjemi. Mode Kbh.* (1939), 75; A. Langseth and B. Bak, *J. Chem. Phys.*, **8**, 403 (1940).

2. O. Hassel and B. Ottar, *Ark. Math. Naturv.*, **45** [10] (1942).
3. (a) B. D. Saksema, *Proc. Indian Acad. Sci., Sect. A*, **12**, 321 (1940); (b) K. W. F. Kohlrausch and H. Wittek, *Z. Phys. Chem.* (Frankfurt am Main), *B*, **48**, 177 (1941).
4. A. Langseth and H. J. Bernstein, *J. Chem. Phys.*, **8**, 410 (1940).
5. O. Hassel and T. Taarland, *Tidsskr. Kjemi Bergv.*, **20**, 167 (1940).

NOTES

[a] Editor's comment: These are the forms now called 'eclipsed' (= *'cis'*) and 'staggered' (= *'trans'*).

[b] Translator's comment: Langseth and Bak[1] actually did not ignore the valence angle strain introduced by a coplanar conformation of the carbon ring. They state, in effect, that if the potential for rotation about the C—C bonds has a minimum corresponding to eclipsing of the H atoms, the classical boat form (C_{2v}) of the molecule will be more stable than the chair form. They point out that such a potential would additionally tend to flatten the ring but that this state could be reached only by overcoming the opposing CCC bond angle strain.

[c] Editor's comment: In the early 1940s, the term 'configuration' included what we now call 'conformation'.

[d] Editor's comment: These terms correspond to 'axial' (= ϵ) and 'equatorial' (= κ) as now used.

[e] Referring to an earlier paper by Hassel (editor's comment).

[f] The interpretation is (nevertheless) quite certain in this case, since the electron diffraction results are supported by dipole moment measurements. If we include consideration of dipole moment measurements, the first alternative for α-benzenehexachloride (p.12) disappears and we are left with the choice between the two last interpretations.

[g] This hypothesis has been superseded: cf. E. L. Eliel, N. L. Allinger, S. J. Angyal and G. A. Morrison, *Conformational Analysis*, Wiley-Interscience, 1965, p. 474.

THE CONFORMATION OF THE STEROID NUCLEUS[1]

D. H. R. BARTON[2]

[from *Experientia* 6, 316–320 (1950)]

In recent years it has become generally accepted that the chair conformation of cyclohexane is appreciably more stable than the boat. In the chair conformation it is possible[3,4] to distinguish two types of carbon–hydrogen bonds; those which lie as in (Ia) perpendicular to a plane containing essentially the six carbon atoms and which are called[3] *polar* (p), and those which lie as in (Ib) approximately in this plane. The latter have been designated[3] *equatorial* (e).[a]

The notable researches of Hassel and his collaborators[5,6] on the electron diffraction of cyclohexane derivatives have thrown considerable light on these more subtle aspects of stereochemistry. Thus it has been shown[5] that monosubstituted cyclohexanes adopt the equatorial conformation (IIa) rather than the polar one (IIb). This is an observation of importance for it indicates that the equatorial conformations are thermodynamically more stable than the polar ones. It should perhaps be pointed out here that although one conformation of a molecule is more stable than other possible conformations, this does *not* mean that the molecule is *compelled* to react as

(Ia)   (Ib)

(IIa)   (IIb)

if it were in this conformation or that it is rigidly fixed in any way. So long as the energy *barriers* between conformations are small, separate conformations cannot be distinguished by the classical methods of stereochemistry. On the other hand a small difference in free-energy content (about one kilocal at room temperature) between two possible conformations will ensure that the molecule appears by physical methods of examination and by thermodynamic considerations to be substantially in only *one* conformation.

The equatorial conformations are also the more stable in both *cis*-1,3- and *trans*-1,4-disubstituted cyclohexanes.[3] Thus *cis*-1,3-dimethylcyclohexane adopts the diequatorial conformation (IIIa) rather than the dipolar one (IIIb), whilst *trans*-1,4-dimethylcyclohexane exists as (IVa) rather than (IVb).

Thermodynamic calculations[3] show that *trans*-1,2-dimethylcyclohexane takes up the diequatorial conformation (V; R = CH$_3$) rather than the dipolar one (VI; R = CH$_3$).

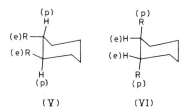

(IIIa)   (IIIb)

(IVa)   (IVb)

For *cis*-1,2-disubstituted cyclohexanes there are two possible conformations. In both of these one of the substituents forms an equatorial bond, the other a polar one. Since these differences in thermodynamic stability between equatorial and polar conformations are presumably of steric origin,[3] it would appear logical to make the larger substituent form the equatorial bond.

(V)   (VI)

Considerations of the same type can be extended to 2-substituted cyclohexanols. Thus[7,8] the *cis* alcohols (VII; R = alkyl), on equilibration by heating with sodium, furnish almost entirely the *trans* isomers (VIII; R = alkyl). In the former one substituent is polar, one equatorial; in the latter both are equatorial. The same conclusion on relative stability is reached from a consideration of thermochemical data.[9] Similarly[10] the 2,6-disubstituted cyclohexanol (IX), with two equatorial and one polar substituents, is isomerized to (X) on equilibration. The situation is the same[8] with the bicyclic *trans*-α-decalol. Here the isomer (XI) is isomerized to (XII) on equilibration.

A consideration of the conformations[11] (XIII) and (XIV), assumed by the steroid nucleus when the A/B ring fusion is respectively *trans* and *cis*, provides a striking illustration of the usefulness of the concept of polar and equatorial bonds. The relationship between the α- and β-nomenclature introduced by Fieser[12] and the occurrence of polar and equatorial bonds is also summarized in (XIII) and (XIV).

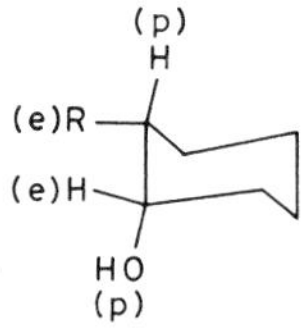
(p)
H
(e)R
(e)H
HO
(p)

(VII)

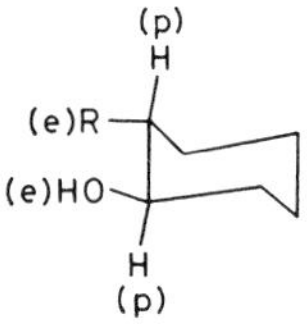
(p)
H
(e)R
(e)HO
H
(p)

(VIII)

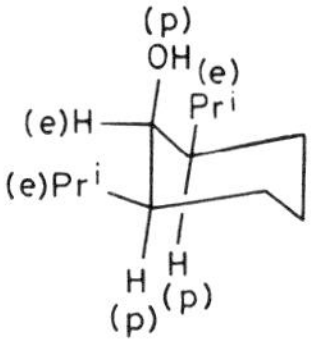
(p)
OH
(e)
Pri
(e)H
(e)Pri
H H
(p) (p)

(IX)

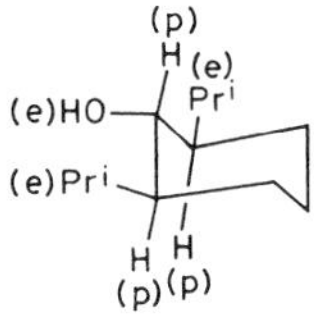
(p)
H
(e)
Pri
(e)HO
(e)Pri
H H
(p) (p)

(X)

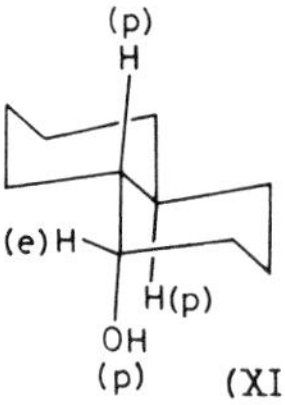
(p)
H
(e)H
H(p)
OH
(p)

(XI)

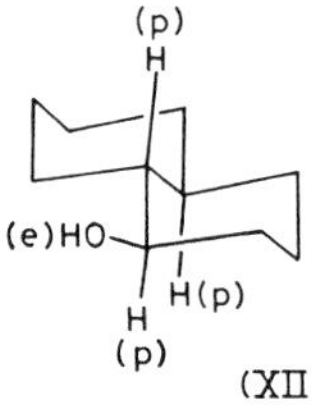
(p)
H
(e)HO
H(p)
H
(p)

(XII)

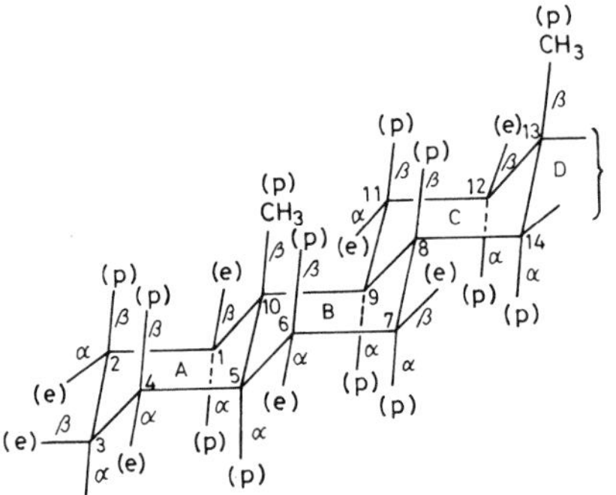
(p)
CH3
A
B
C
D

(XIII)

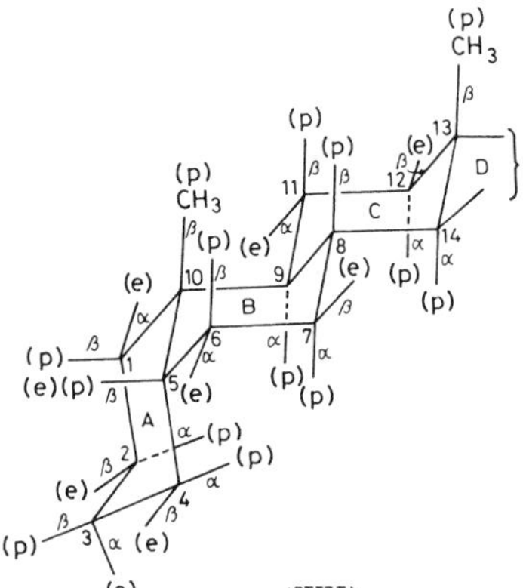
(p)
CH3
A
B
C
D

(XIV)

Thermodynamic considerations. In a number of cases equilibration of hydroxyl groups at secondary positions in the steroid nucleus has been carried out. At other positions the corresponding ketones have been reduced by sodium and alcohol, a process which (in cyclohexane derivatives) is well established to give the thermodynamically more stable alcohols in approximately the same proportions as from equilibration experiments.[7,8] It is possible therefore to see how well the concept of more stable equatorial conformations is obeyed. As set out in Table I the expected relationships are observed. Also included in this table is a reference to the equilibration of 5α, 6β-dibromocholestane with the $5\beta,6\alpha$-isomer, for this is clearly relevant to the issue under discussion.

TABLE I

Observation	Exptl. method	References
Cholestane series		
2α (e) more stable than 2β (p)	Reduction of 2-one	L. Ružička, P. A. Plattner and M. Furrer, *Helv. Chim. Acta*, **27**, 524 (1944).
3β (e) more stable than 3α (p)	Equilibration	pp. 98, 636[a]
4α (e) more stable than 4β (p)	Reduction of 4-one[b]	R. Tschesche and A. Hagedorn, *Ber.*, **68**, 2247 (1935); L. Ružička, P. A. Plattner and M. Furrer, *loc. cit.*
6α (e) more stable than 6β (p)	Reduction of 6-one	pp. 223, 653[a]
7β (e) more stable than 7α (p)	Reduction of 7-one[c]	I. M. Heilbron, W. Shaw and F. S. Spring, *Rec. Trav. Chim.*, **57**, 529 (1938).
5β (e, p), 6α (e)-dibromide more stable than 5α (p), 6β (p)-dibromide	Equilibration	D. H. R. Barton and E. Miller, *J. Amer. Chem. Soc.*, **72**, 1066 (1950).
Coprostane series		
3α (e) more stable than 3β (p)	Equilibration	pp. 99, 636[a]
11α (e) more stable than 11β (p)	Equilibration	p. 408[a]
12β (e) more stable than 12α (p)	Equilibration	pp. 461, 657[a]

[a]   All references reported in this way are to L. F. Fieser and M. Fieser, *Natural Products Related to Phenanthrene*, 3rd ed., Reinhold, 1949.

[b]   The configurations are assigned (*vide infra*).

[c]   According to the standard tables of D. H. R. Barton and W. Klyne, *Chem. Ind.* (London), 1948, 755, 7β-hydroxycholestane should have $[\alpha]_D$ *ca.* $+52°$, whilst the 7α-isomer should exhibit $[\alpha]_D$ *ca.* $+8°$. I. M. Heilbron, W. Shaw and F. S. Spring, *loc. cit.*, observed $[\alpha]_D$ $+51°$ and therefore the configuration of their alcohol must be 7β.

Elimination evidence. Reactions whose mechanisms require concerted 1,2-elimination should proceed more readily when the four centres involved (the two carbon atoms and the two substituents) lie in one plane. For concerted ionic elimination reactions in cyclohexane derivatives the optimum arrangement of the substituents for the minimization of the activation energy is that in which both are polar.[13,14] There is much evidence in the literature which confirms this. Thus[7] *cis*-2-substituted cyclohexanols (VII) undergo acid-catalysed dehydration [elimination of H (p) and OH (p)] more readily than the *trans* isomers (VIII). In the menthol series[17] neomenthol (XV; R = CH_3, R' = H) loses water easily relative to menthol (XVI; R = CH_3, R' = H) and neoisomenthol (XV; R = H, R' = CH_3) dehydrates easily relative to isomenthol (XVI; R = H, R' = CH_3).[18] There are a number of interesting examples of this sort of phenomenon in steroid compounds. A summary is given in Table II.

(XV) (XVI)

TABLE II

Observation easy elimination of	References
Cholestane series	
6β-OH (p) and 5α-H (p)	D. H. R. Barton and W. J. Rosenfelder, *J. Chem. Soc.,* 1949, 2459.
6β-H (p) and 5α-Cl (p)	D. H. R. Barton and E. Miller, *J. Amer. Chem. Soc., 72,* 1066 (1950).
6β-Br (p) and 5α-Br (p) 7α-OH (p) and 8β-H (p)	D. H. R. Barton and E. Miller, *loc. cit.* pp. 241, 242, 631[a]
Coprostane series	
7α-OH (p) and 8β-H (p)	pp. 118, 631[a]
11β-OH (p) and 9α-H (p)	pp. 408, 630[a]
11β-Br (p) and 9α-H (p)	pp. 460, 631[a]

[a] All references reported in this way are to L. F. Fieser and M. Fieser, *Natural Products Related to Phenanthrene*, 3rd ed., Reinhold, 1949.

Steric hindrance evidence. The applicability of steric hindrance evidence in the assignment of configuration has long been recognized, although such assignments are not always reliable.[19] It seems possible to explain the relative magnitudes of many of the phenomena of steric hindrance in cyclohexane derivatives on the basis that polar bonds are more hindered than the corresponding equatorial bonds. An inspection of models makes this reasonable for a polar bond is always close in space to two other polar bonds each attached to the next but one carbon atom, whereas there is no similar relationship for equatorial bonds.

In support[7] of this generalization it has been observed that *cis*-2-substituted cyclohexanols (VII) with polar hydroxyls are more difficult to esterify, and their esters more difficult to hydrolyse, than the corresponding *trans* alcohols and their esters. The same effects are observed with *trans*-α-decalol.[20] The esters of the alcohol (XI) (polar hydroxyl) are more difficult to hydrolyse than those of the alcohol (XII) (equatorial hydroxyl). In the menthol series[17] menthol (XVI; R = CH_3, R' = H) is more easily esterified than neomenthol (XV; R = CH_3, R' = H) and a similar relationship holds for isomenthol (XVI; R = H, R' = CH_3) and neoisomenthol (XV; R = H, R' = CH_3).

However, a reverse relationship holds[7] for chromic acid oxidation of 2-substituted cyclohexanols. Here the *cis* alcohols are oxidized more rapidly than the *trans*. This observation is adequately accommodated by the present theory if the rate-determining step is attack upon the carbon—hydrogen bond rather than upon the carbon—hydroxyl linkage.[b]

The situation in the steroid field is summarized in Table III. In every case the expected order of hindrance holds good. Also included are data for oxidations of alcohols by Br^+ to give the corresponding ketones. If such oxidations are assumed to involve attack upon the carbon—hydrogen bond then the results are in agreement with the other observations summarized in the table.

Although the concept of polar and equatorial bonds is not, of course, applicable to cyclopentane, it is of interest to note that the 17α-bond in the steroid nucleus has, because of the ring fusion to a six-membered ring, the character of a polar bond with respect to that ring. Also the 17β-bond has in its relationship to ring C the aspect of an equatorial bond. These facts are in agreement with the greater thermodynamic stability of 17β-substituents and the greater degree of steric hindrance shown by 17α-substituents.[21]

TABLE III

Observation	References
Cholestane series	
2β-OH (p) more hindered than 2α-OH (e)	A. Fürst and P. A. Plattner, *Helv. Chim. Acta*, **32**, 275 (1949).
3α-OH (p) more hindered than 3β-OH (e)	pp. 635, 636[a]
6β-OH (p) more hindered than 6α-OH (e)	p. 223[a]
6α-H (e) more easily oxidized than 3α-H (p)	L. F. Fieser and S. Rajagopalan, *J. Amer. Chem. Soc.*, **71**, 3938 (1949).
3β-H (e) more easily oxidized than 3α-H (p)	G. Vavon and B. Jacubowicz, *Bull. Soc. Chim. Fr.*, **53** [4], 581 (1933).
Coprostane series	
3β-OH (p) more hindered than 3α-OH (e)	pp. 635, 636[a]
6β-OH (p) more hindered than 6α-OH (e)	p. 652[a]
11β-OH (p) more hindered than 11α-OH (e)	p. 408[a]
12α-OH (p) more hindered than 12β-OH (e)	p. 658[a]
7α-OH (p) and 12α-OH (p) more hindered than 3α-OH (e)	p. 125[a]
7β-H (e) and 12β-H (e) more easily oxidized than 3β-H (p)	p. 126[a]; L. F. Fieser and S. Rajagopalan, *J. Amer. Chem. Soc.*, **71**, 3935 (1949).

[a] All references reported in this way are to L. F. Fieser and M. Fieser, *Natural Products Related to Phenanthrene*, 3rd ed., Reinhold, 1949.

Use of the concept. It will be clear that it is possible to assign configurations on the basis of the concept of polar and equatorial bonds. One such example has already been given in Table I. An additional illustration is provided by *trans*-β-decalol.[8,19] The more stable epimer m.p. 75° must have the hydroxyl in the equatorial conformation as in (XVII); this is in agreement with the fact that its esters are more rapidly hydrolysed than those of the epimeric (polar hydroxyl) alcohol. Other examples are mentioned below.

Extension to di- and triterpenoids. It would seem reasonable to extend the concept of equatorial and polar bonds to the correlation of the stereochemistry of other ring systems built up from fused cyclohexane units. Thus ring A of the diterpenoid abietic acid may be represented[22] by (XVIII; R = CO_2H, R' = CH_3) with the carboxyl occupying an equatorial conformation. It is understandable then that the esters of this acid should be more easily hydrolysed than those of (say) podocarpic acid where ring A is as shown in (XVIII; R = CH_3, R' = CO_2H), for in the latter the carboxyl occupies the more hindered polar conformation.

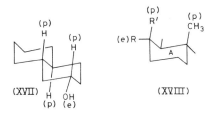

(XVII)

(XVIII)

Now that it is recognized[22] that rings A and B of the α- and β-amyrin groups of triterpenoids and also[23] those of the lupeol group are *trans* fused, it is possible to make a tentative representation of their stereochemistry as shown in (XIX; R = H). Placing the hydroxyl in the equatorial conformation explains the more facile hydrolysis of β-amyrin acetate relative to epi-β-amyrin acetate[24] and of lupanol relative to epi-lupanol.[25] It also accounts for the easy elimination of water accompanied by molecular rearrangement, which is induced in these compounds or their derivatives by treatment with phosphorus pentachloride.[26] Such a reaction then becomes comparable to the very easy dehydration of isoborneol to give camphene, in that all the four atomic centres of importance in the reaction lie in one plane. The marked hindrance of the 7-hydroxyl group in sumaresinolic acid and its easy elimination under acid dehydrating conditions[27] are best explained if it has the polar conformation as in the part expression (XIX; R = OH).

(XIX)

In connection with the nomenclature of triterpenoids it would appear desirable to extend Fieser's α-, β-convention for steroids to cover triterpenoid stereochemistry also. A convenient reference point is the C_5 methyl group. Substituents on the same side of the main-ring plane as this methyl group should be regarded as having the β-configuration, those on the opposite side as having the α-configuration. Thus sumaresinolic acid would be designated $2\beta,7\beta$-dihydroxyolean-12-ene-17-carboxylic acid.

REFERENCES

1. The word 'conformation' is used to denote differing strainless arrangements in space of a set of bonded atoms. In accordance with the tenets of classical stereochemistry, these arrangements represent only one molecular species.
2. Harvard University Visiting Lecturer, 1949–50, Harvard University, Cambridge 38, Mass.
3. C. W. Beckett, K. S. Pitzer and R. Spitzer, *J. Amer. Chem. Soc.,* **69**, 2488 (1947).
4. O. Hassel's nomenclature (5) is different, but the distinction remains the same.
5. O. Hassel and H. Viervoll, *Acta Chem. Scand.,* **1**, 149 (1947).
6. See O. Hassel and B. Ottar, *Acta Chem. Scand.,* **1**, 929 (1947), for a summarizing paper and references to earlier work.
7. G. Vavon, *Bull. Soc. Chim. Fr.,* **49**, 937 (1931).
8. W. Hückel, *Ann. Chem.,* **533**, 1 (1937).
9. A. Skita and W. Faust, *Ber. Deut. Chem. Ges.,* **64**, 2878 (1931).
10. G. Vavon and P. Anziani, *Bull. Soc. Chim. Fr.,* **4** [5], 1080 (1937). In connection with the conformations of polysubstituted cyclohexanes it should be mentioned that O. Bastiansen, O. Ellersen and O. Hassel, *Acta Chem. Scand.,* **3**, 918 (1949), have recently shown that the five stereoisomeric benzene hexachlorides assume, in agreement with our general argument, those conformations which have the maximum possible number of equatorial carbon–chlorine bonds.
11. Conformations (XIII) and (XIV) are unambiguous representations of the steroid nucleus provided that rings A, B and C are chairs. This is almost certainly true for a *trans*-A/B ring fusion (compare the X-ray evidence of C. H. Carlisle and D. Crowfoot, *Proc. Roy. Soc., Ser. A,* **184**, 64 (1945) on the conformation of cholesteryl iodide) and a similar situation, at least in solution, probably holds for a *cis*-A/B fusion. The justification for the latter has been more extensively presented elsewhere (14, 15). See also the discussion by H. Sobotka, *The Chemistry of the Steroids,* Williams and Wilkins, 1938, pp. 48ff.
12. L. F. Fieser, *The Chemistry of Natural Products Related to Phenanthrene*, 1st ed., Reinhold, New York, 1936.
13. E. D. Hughes and C. K. Ingold *et al., J. Chem. Soc.,* 2117 (1948).
14. D. H. R. Barton and E. Miller, *J. Amer. Chem. Soc.,* **72**, 1066 (1950).
15. O. Bastiansen and O. Hassel, *Nature,* **157**, 765 (1946); D. H. R. Barton, *J. Chem. Soc.,* 340 (1948).
16. See ref. 5 and papers there cited.
17. For summary see J. L. Simonsen and L. N. Owen, *The Terpenes,* Vol. I, Cambridge University Press, 1947.
18. Of course for pyrolytic elimination of substituents by 'unimolecular' mechanisms (see D. H. R. Barton, *J. Chem. Soc.,* **1949**, 2174) *cis*-elimination is the rule and the discussion given here is no longer relevant.
19. See ref. 8. Compare reference 21 in which L. F. Fieser has discussed steric effects under the headings intraradial and extraradial.
20. W. Hückel *et al., Ann. Chem.,* **533**, 128 (1937).
21. L. F. Fieser, *Experientia,* **6**, 312 (1950).
22. D. H. R. Barton, *Quart. Rev.* (London), **3**, 36 (1949).
23. T. R. Ames and E. R. H. Jones, *Nature,* **164**, 1090 (1949).
24. L. Ružička and H. Gubser, *Helv. Chim. Acta,* **28**, 1054 (1945); these authors assigned the opposite configuration at C_2.

25. R. Nowak, O. Jeger and L. Ružička, *Helv. Chim. Acta,* **32**, 323 (1949). The equatorial conformation for the hydroxyl group in these compounds is also indicated by the fact that β-amyrin is more stable thermodynamically than epi-β-amyrin (L. Ružička and W. Wirz, *ibid.,* **24**, 248 (1941)).
26. L. Ružička, M. Montavon and O. Jeger, *Helv. Chim. Acta,* **31**, 819 (1948); and earlier papers from the same laboratory.
27. L. Ružička, O. Jeger, A. Grob and H. Hösli, *Helv. Chim. Acta,* **26**, 2283 (1943).

NOTES

[a] Editor's comment: The word 'axial' is currently used to mean the same thing as the word 'polar' in this paper.

[b] Editor's comment: This explanation has been superseded: cf. J. Schreiber and A. Eschenmoser, *Helv. Chim. Acta,* **38**, 1529 (1955).

NEWER DEVELOPMENTS OF THE CHEMISTRY OF
MANY-MEMBERED RING COMPOUNDS

(Centenary Lecture delivered before the Chemical Society in London on
February 17th, 1949)

V. PRELOG

[from *Journal of the Chemical Society* 420–428 (1950)]

I am able to speak about the newer developments of the chemistry of macrocyclic compounds thanks to the fortunate circumstance that I am working in the Laboratory of organic chemistry at the Institute of Technology in Zürich. In this laboratory the many-membered rings were discovered by L. Ružička and the fundamental analytical and synthetical work has been done on this interesting group of compounds. Ružička himself lectured on his work before the Chemical Society in London almost exactly 15 years ago.[1] The most important theoretical consequences of these investigations and especially their influence on the so-called strain theory form part of general text-book chemistry to-day. Moreover, the perfume industry could not dispense with the artificial musk-like scents, which became available through this work.

In spite of the wide extent of this field only few laboratories have devoted them-selves to the chemistry of many-membered ring compounds. I would like to mention, in addition to the Zürich Laboratory and a connected group led by M. Stoll in Geneva, Carothers, Adams and Blomquist in America, and Ziegler, Lüttringhaus and Huns-diecker in Germany. If I concern myself mostly with the newer work of our own group, it is only on account of restricted time.[2] For the same reason I cannot deal now with the very interesting recent work on many-membered ring peptides, antibiotics and alkaloids.

The reason why the chemistry of many-membered ring compounds was not more intensively studied lay in their difficult accessibility. In the heroic age, Ružička and his co-workers prepared the many-membered ring ketones, which they used as starting materials for other compounds of this group, by dry distillation of the rare-earth salts of dicarboxylic acids.[3] In favourable cases the yields reached a few per cent. The introduction of the dilution principle by K. Ziegler, H. Eberle and H. Ohlinger[4] did not make the many-membered ring derivatives readily available, since the technique of ring-closure in great dilution is rarely simple. All general procedures for the prepara-tion of macrocyclic compounds had still another important disadvantage. The com-pounds of medium ring-size, containing 8 to 12 members, could be prepared only in disappointingly low yields of a few tenths per cent.

The introduction of the very advantageous acyloin synthesis,[5] which allows pre-paration in very good yields of ring compounds having more than 8 members, formed therefore a starting point for new approaches to this field. This synthesis leads from easily accessible dicarboxylic acid esters in a simple way without using great dilutions to macrocyclic acyloins, which can be converted easily into various other cyclic compounds. The following formulae show a variety of reactions, which were carried out recently:

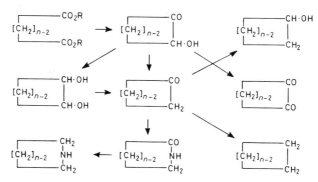

Fig. 1 illustrates the dependence of yield on ring size for the three most important general procedures for the preparation of many-membered ring compounds: the preparation of ring ketones according to Ružička, the preparation of imino-nitriles by Ziegler's method, and the acyloin procedure. One can see that it is now possible to obtain ring-compounds of every size in reasonable yield. In this way the field of many-membered ring-compounds became a real playground for the organic chemist. The acyloin procedure leads to the democratisation of this hitherto very aristocratic group of compounds. It is now no more difficult to obtain *cyclo*-decanone than any higher aliphatic ketone.

A very elegant application of the acyloin synthesis to a macrocyclic compound is exemplified in the synthesis of civetone by M. Stoll, J. Hulstkamp and A. Rouvé.[6] It was possible to synthesise both *cis*- and *trans*-isomers of civetone and this is the first case of a double bond incorporated in a ring in both *cis*- and *trans*-configuration in known fashion.

The problem of the surprisingly good yields of many-membered ring acyloins obtained from the acyloin synthesis in concentrated solutions is explained by our reaction mechanism shown schematically in Fig. 2.[7] The two electrophilic carbon

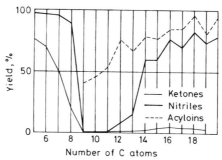

Figure 1.

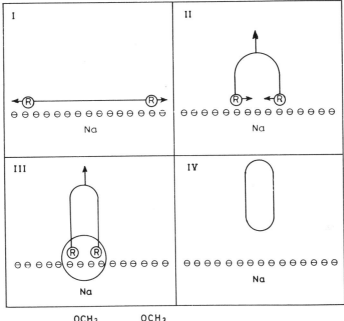

Figure 2.

atoms at the ends of the chain of a dicarboxylic acid ester are first adsorbed by the electron-covered surface of the molten sodium. So far as the flexibility of the carbon chain allows, the electrophilic residues can slide over the metal surface and can thus approach each other. As less energy is necessary for this than for splitting the molecule off the surface, the collisions of the adsorbed molecule with other molecules lead to the approach of the terminal carbon atoms and finally to the ring closure. After the ring closure the molecule no longer possesses electrophilic centres and is therefore no longer bound to the surface.

The work on many-membered rings was from the beginning carried out with two motives, the one practical — the preparation of musk-like perfumes — and the other theoretical — the investigation of the influence of ring size on the physical and chemical properties. It seemed that this second aim would not have much interest since according to older views the higher unstrained ring homologues of *cyclo*hexane should not differ from it more than the higher aliphatic homologues differ from the lower. In fact *the physical and the chemical properties of the many-membered ring compounds do show a peculiar and unexpected dependence on ring size.*

Fig. 3 shows, for example, that the curves of melting point against ring size do not rise steadily as with aliphatic homologues.[8] As no simple explanation of these curves is

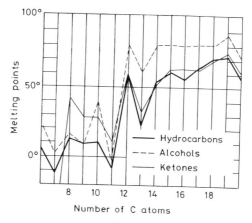

Figure 3.

possible we shall turn rather to consideration of those more readily explicable properties which are not dependent on lattice forces. Such, for example, are the density and
the molecular refraction in the liquid state. Most of the investigated series of alicyclic
many-membered ring compounds show a maximum of density and a depression of
molecular refraction at medium ring size. In this manner they differ from the aliphatic
homologous series, which show a steady increase or decrease of density and no
appreciable deviation $E\Sigma$ from the calculated molecular refraction. Fig. 4 describes
the above-mentioned properties as a function of ring size in the series of cyclanol
acetates.[8] Cyclanol acetates are specially suitable for such measurements, because the
whole series is liquid at room temperature so that no corrections are necessary.

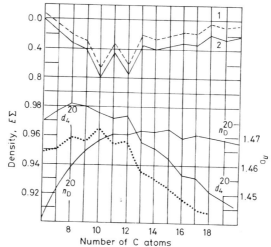

Figure 4. Atomic refractions: curve 1, Eisenlohr; curve 2, Vogel.

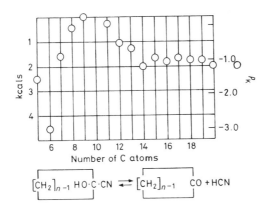

Figure 5.

To the extreme values of the physical properties correspond extreme values of the chemical properties of the medium-size ring compounds. As the first example of the chemical properties, the dependence of the equilibrium constants on ring size is shown in Fig. 5 for the reaction between cyclanones and hydrogen cyanide in alcoholic solution.[9] The negative logarithms of the dissociation constants and the free energy of the reaction are plotted against ring size, the influence of which is unexpectedly great. The two extreme values lie on the limits of possible experimental determination. That means that *cyclo*hexanone cyanohydrin scarcely dissociates at all, whereas *cyclo*decanone does not add on hydrogen cyanide. For comparison, the dissociation constant for dioctyl ketone cyanohydrin is given on the right-hand side of the graph.

As a further example of the dependence of an equilibrium constant on the ring size we can consider the dissociation constants of polymethylene-imines shown in Fig. 6.[10] We also measured polarographically the half-wave potentials for the cathodic reduction of the Girard-T derivatives of cyclic ketones, which are shown in Fig. 7.[11] In spite of the fact that it is not a reversible reaction, we obtained a shape of curve which shows a dependence of half-wave potentials on ring size very similar to those in the two foregoing examples.

The common feature in all these reactions is that at medium-ring size that reaction component is stabilised which contains a more nucleophilic oxygen or nitrogen: for

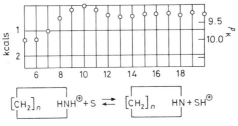

Figure 6.

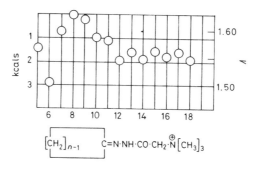

Figure 7.

example, ketone is stabilised with respect to cyanohydrin, imine with respect to the corresponding ammonium salt, hydrazone with respect to the reduced form. One possible explanation for this fact is that the nucleophilic centres interact with hydrogens of the polymethylene chain through formation of some kind of intramolecular hydrogen bridge, which is particularly strong if the ring has a medium size. The formation of such intramolecular hydrogen bridges is only possible if we assume a certain constellation[a] of the poly-membered ring. As constellation we define those forms of the molecules which result from free rotation around single bonds, for example, the chair and the boat form of *cyclo*hexane. With respect to stability of form, the poly-membered rings stand in between the rigid small rings and the mobile aliphatic open chains, and therefore these compounds provide an interesting material for studies of the influence of constellation on chemical and physical properties. The following two factors will determine mainly the energy and the probability of a constellation: first,

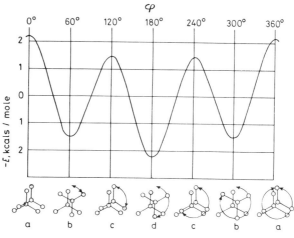

Figure 8.

the van der Waals radii of the atoms, which are well illustrated by Stuart models, and secondly, such interactions as are not shown by atomic models — that is, forces not expressed in classical formulae, for instance, those which restrict rotation around a single bond in aliphatic compounds in contradiction to the views of classical stereo-chemistry.[12] These forces cause the inequality of various constellations of a poly-methylene chain from the energetics standpoint. The changes of the potential energy of a chain of four methylene groups around the central C—C bond can be described approximately by the formula[13]

$$E = -1·7 \cos \phi - 0·5 \cos \phi \text{ kcals per mole}$$

This function is shown together with corresponding configurations in Fig. 8.

Two constellations b and d, which are separated from each other by energy barriers, are especially favoured. In a normal paraffin hydrocarbon the constellation d, which corresponds to the zigzag form of the chain, predominates. In cyclo-paraffins the poly-methylene chain deviates the more from the constellation d the smaller the ring. In cyclopentane and its lower ring homologues the chain is forced by ring strain into the energetically unfavourable constellation a. cycloPentane itself has, according to K. S. Pitzer,[14] a non-planar form because the deformation of the tetrahedral valency-angles of the non-planar form is less serious than the unfavourable constellation of the planar form. From this it is understandable that the chair form, which is built from 6 energetically favourable constellations, predominates in cyclohexane over the boat form, in which constellation b occurs 4 times and the unfavourable constellation a twice.

These considerations being transferred to homologous cyclo-paraffins and simulta-neous account being taken of van der Waals radii, one has the interesting result that medium-sized rings with 8—12 members can only be built from less favourable con-stellations. In contrast to this, the chains of the higher cyclo-paraffins can again be constructed from energetically favourable constellations b and d. Just as in small rings classical strain and constellation work against each other, so in medium-sized rings van der Waals repulsive forces and constellation are opposed. The medium-sized rings suffer a greater interaction between atoms than in cyclo-hexane: they are therefore at a higher energy level than unbranched aliphatic compounds.

Maxima and minima in the physical properties of medium-sized ring compounds may be explained in this way, for example, the reduction of electron polarisability and consequently of molecular refraction by the stronger interaction of non-adjacent atoms in the medium-sized rings.

In considering the influence of ring size on the chemical properties, for example, on the dissociation constants of the cyanohydrins, it should be realised that in small-ring ketones the oxygen stands outside the ring. In medium-sized rings we have many possible forms in between the two extreme constellations, which may be termed 'O-outside' and 'O-inside'. Figs. 9 and 10 show such extreme constellations for cyclo-decanone. The O-outside constellation of medium-sized ring ketones are marked out, first, by a more favourable constellation of the polymethylene chain,[b] and secondly, by an intramolecular interaction between the carbonyl oxygen and the hydrogens of the polymethylene chain; the distance between these atoms is especially small in medium-sized ring ketones. The interaction is stronger in the ketone than in the

Figure 9.

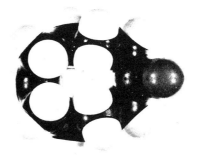

Figure 10.

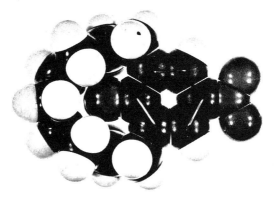

Figure 11.

cyanohydrin and stabilises the former, as was found experimentally. Very similar considerations can be developed for the other reactions mentioned.

In an attempt to prove the consequences of the 'O-inside' configuration of cyclanones, Dr Günthard studied the infra-red absorption spectra. As expected, the frequency of C=O bond-stretching band was smaller with a medium-ring ketone than in *cyclo*hexanone or a large-ring ketone. The following values (in cm^{-1}) were obtained in the liquid state: *cyclo*hexanone 1710, *cyclo*octanone 1692, *cyclo*nonanone 1698, *cyclo*decanone 1694, *cyclo*undecanone 1700, *cyclo*dodecanone 1697, *cyclo*trideca-none 1707.^{c}

In addition to the physical and physicochemical methods of approach, we tried to examine the question of the constellation of many-membered ring compounds by the classical methods of organic chemistry. Assuming that the rate of ring closure is influenced by constellation, it should be possible under favourable conditions to obtain information about the constellation of a monocyclic compound by transforming it into a bicyclic compound. We carried out a series of such experiments and found some facts which deserve brief mention.

Robinson and his co-workers[15] showed that condensation of the *cyclo*hexanone-carboxylic esters with quaternary bases of dimethylaminobutanone ('Mannich base') gives compounds, which can be simultaneously saponified, decarboxylated, and cyclised to $\alpha\beta$-unsaturated octalones. We obtained analogous products (II) by condensing the cyclic β-keto-carboxylic esters (I) containing a 7- or an 8-membered ring.[16] In the case of the 8-membered ring we isolated a second isomeric $\alpha\beta$-unsaturated ketone (III), in which the carbonyl of the side chain had condensed with the active methylene group in the ring. The two isomers could be separated easily by making use of the fact that the second of them does not give the carbonyl group reactions, for example, does not react with Girard reagent-T. Ultra-violet absorption spectra and Kuhn—Roth oxidation provide evidence for the constitution. The compound with the bicyclic system condensed in the 1,2-position has an absorption maximum at a shorter wave-length and gives no acetic acid by oxidation, whereas the isomer with the bicyclic system condensed in the 1,3-position gives one molecule of acetic acid by oxidation and absorbs at longer wave-length. Using higher homologues of *cyclo*octanonecarboxylic ester, we got only products of the type (III).

These experiments permit some definition of the limits of validity of Bredt's rule, which states that in bridged-ring systems a double bond cannot start from a bridgehead. Fundamentally, Bredt's rule expresses the requirement of effective overlap of the p-electron orbitals constituting the π-electron orbitals of the double bond. Twisting of the inter-atomic axis in a bridged-ring system is thus the simplest case of that steric hindrance of π-electron resonance which has recently been much investigated in systems containing several conjugated double bonds. Bredt's rule is formally broken by the products (III) of the Robinson reaction containing a 1,3-bicyclic system, and it is therefore clear that a double bond may lie at the bridgehead of a bicyclic system if the ring only be large enough.

We investigated this problem more thoroughly by a variant of Robinson's synthesis.[17] We condensed the cyclanone carboxylic esters with γ-chlorocrotyl chloride (1,3-dichlorobut-2-ene) instead of with the quaternary salt of the 'Mannich-base' and treated the products (IV) with concentrated sulphuric acid according to O. Wichterle.[18] The γ-chlorocrotyl derivatives give under these conditions γ-ketobutyl derivatives, which react further as in Robinson's synthesis but without saponification and decarboxylation. The carbonyl of the β-keto-carboxylic ester is less reactive than in ketones, and only bicyclic compounds condensed in the 1,3 position are formed independently of ring size. The compounds (V) from the starting material with 6- or 7-membered rings contain, however, no $\alpha\beta$-unsaturated carbonyl, as shown by ultra-violet absorption spectra, or in other words they obey Bredt's rule. In contrast, the compounds (VI) with 8-membered and higher ring are $\alpha\beta$-unsaturated ketones. Thus the limit of applicability of Bredt's rule lies between the systems with a 7- and an 8-membered ring.

The bicyclic α-keto-carboxylic acids (V) and (VI), which are obtained by the variant of the Robinson synthesis, show an interesting behaviour on decarboxylation. The compound (V) with two 6-membered rings cannot be decarboxylated even under drastic conditions. Similar examples are known from terpene chemistry, for example, camphoronic acid. On the other hand, the analogous acids, (V) with a 7- and (VI) with an 8-membered ring, decarboxylate easily at $240°$, whereas the acid (VI) with a 13-membered ring decarboxylates even during the alkaline saponification of the ester. If we assume that an anion at the bridgehead is an intermediate, which must be stabilised by resonance with the carbonyl for the activation energy to be sufficiently low, then we can use these results as a contribution to the knowledge of the validity of Bredt's rule.

I should like to mention that γ-chlorocrotyl carboxylic esters (IV) can be saponified and simultaneously decarboxylated with a mixture of hydrochloric and acetic acids and that treatment of the resulting α-chlorocrotyl ketones (VIII) with sulphuric acid gives the same products (II) as the Robinson synthesis.

Since all experiments showed that the carbonyl in poly-membered ring compounds is feebly reactive, whereas the α-methylene groups are strongly reactive, we tried to condense cyclic ketones with nitromalonic dialdehyde (IX), which gives substituted

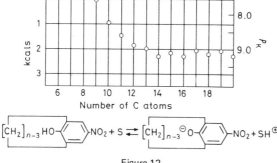

The reaction scheme at the top of the page shows compounds (IX), (X), (XI), (XIII), (XII), (XVI), (XV), and (XIV) with the following transformations:

$$[CH_2]_{n-3} \begin{array}{c} CH_2 \\ CO \\ CH_2 \end{array} + \begin{array}{c} OCH \\ CH \cdot NO_2 \\ OCH \end{array} \longrightarrow [CH_2]_{n-3} \begin{array}{c} C = CH \\ CO \quad {}^{\ominus}C \cdot NO_2 \\ CH - CH \cdot OH \end{array} \xrightarrow{10-20} [CH_2]_{n-3} \; {}^{\ominus}O - \langle \rangle - NO_2$$

(IX)   (X)   (XI)

$\downarrow$ 6,7

θ

$\downarrow\uparrow$ 8 (X) $\downarrow\uparrow$ 9-20 (XI)

9

$[CH_2]_{n-3} \; O = \langle \rangle - NO_2$ (XIII) $[CH_2]_{n-3} \; HO - \langle \rangle - NO_2$ (XII)

$[CH_2]_{n-3} \; HO - \langle \rangle - OH$ (XVI) $\leftrightarrows$ $[CH_2]_{n-3} \; O = \langle \rangle = O$ (XV) $\leftarrow$ $[CH_2]_{n-3} \; HO - \langle \rangle - NH_2$ (XIV)

p-nitrophenols easily with aliphatic ketones as was shown by Hill and by Kenner.[19] In fact, the many-membered ring ketones with a more than 8-membered ring gave compounds of the expected formulae.[20] However, the influence of ring size on the formation of the bicyclic system showed an interesting effect. With 10- and higher-membered ring ketones yellow p-nitrophenoxides (XI) were directly obtained, which yielded on acidification p-nitrophenols (XII). Ketones with an 8- and a 9-membered ring gave instead colourless condensation products, which contained one molecule of water more than the expected p-nitrophenoxides. These products differ in their absorption spectrum from the p-nitrophenoxides and probably have the constitution (X).[21] On acidification, the product with a 9-membered ring gives immediately the corresponding p-nitrophenol, which can be converted by alkali into the normal p-nitrophenoxide. In contrast to this, the primary condensation product which contains an 8-membered ring gives on acidification a compound (XIII) which is tautomeric with

$$[CH_2]_{n-3} \; HO - \langle \rangle - NO_2 + S \rightleftarrows [CH_2]_{n-3} \; {}^{\ominus}O - \langle \rangle - NO_2 + SH^{\oplus}$$

Figure 12.

the expected p-nitrophenol, as could be shown by a thorough investigation. This compound is converted by alkali back into the colourless salt (X), which contains 1 molecule of water more than the p-nitrophenoxide. We have here an interesting case of a compound in which the ring strain has overcome the tendency to form an aromatic system. This is in accordance with the model, which shows that the compound (XIII) is much less strained than the m-bridged benzene derivative. The m-bridged p-nitrophenols can serve as starting material for other series of m-bridged benzene derivatives, for example, 4-aminophenols (XIV), p-benzoquinones (XV), and quinols (XVI).[22]

In the above mentioned m-bridged bicyclic compounds the many-membered rings have forms which are very similar to the forms of monocyclic ketones with 'O-inside' constellations: this can be seen from the model of a m-bridged p-nitrophenol, as shown in Fig. 11. Therefore it was interesting to test whether the influence of the medium-sized rings on the chemical properties of the bicyclic compounds was the same as in the monocyclic compounds. The aromatic part of the molecule served in these cases as a kind of indicator for the forces inside the ring.

We measured the dissociation constants of the m-bridged p-nitrophenols in 80% Methylcellosolve[21] (cf. Fig. 12) and the redox potential of the m-bridged p-benzoquinones[23] (cf. Fig. 13). One can see in both cases that the medium-sized rings stabilise the reaction component with the more nucleophilic oxygen, as would be expected on the assumption of the formation of an intramolecular hydrogen bridge.

I believe that the investigations which we have carried out show clearly that a *medium-size ring effect exists which was not predicted by classical chemistry.* I hope that our interpretation of this effect as a consequence of the constellation and intramolecular interactions of atoms across space is not very far from the truth, and that it can be used as a basis for a more exact theory and as a stimulus for further experiments.

The constellation has hitherto often been neglected as a very important factor which influences the reaction equilibria and reaction rates. The advances in physical approach to this problem will perhaps allow the organic chemist to make in the future more valuable contributions from his standpoint than was possible before.

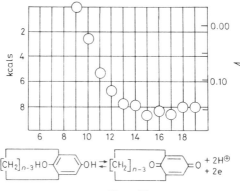

Figure 13.

In conclusion, my best thanks for their wholehearted co-operation are due to my younger colleagues P. Barman, L. Frenkiel, O. Häfliger, H. Günthard, W. Ingold, Margrit Kobelt, K. Wiesner, M. M. Wirth and M. Zimmermann, who performed the laborious and, in experimental details, not always very interesting work. My thanks are due also to Dr G. W. Kenner for help in the translation of the manuscript.

REFERENCES

1. *Chem. and Ind.,* 1935, **54**, 2.
2. Cf. reviews of newer work on many-membered ring compounds by M. Stoll, *Chimia,* 1948, **2**, 217, and M. G. J. Beets, *Chem. Weekblad,* 1948, **44**, 297.
3. L. Ružička, W. Brugger, M. Peiffer, H. Schinz and M. Stoll, *Helv. Chim. Acta,* 1926, **9**, 499.
4. *Annalen,* 1933, **504**, 94.
5. V. L. Hansley, U.S.A. Pat. 2,226,268; V. Prelog, L. Frenkiel, Margrit Kobelt and P. Barman, *Helv. Chim. Acta,* 1947, **30**, 1741; M. Stoll and J. Hulstkamp, *ibid.,* pp. 1815, 1837.
6. *Ibid.,* 1948, **31**, 543.
7. *Ibid.,* 1947, **30**, 1742.
8. Cf. L. Ružička, M. Stoll, H. W. Huyser and H. A. Boekenoogen, *ibid.,* 1930, **13**, 1152; L. Ružička and G. Giacomello, *ibid.,* 1937, **20**, 548; L. Ružička, Pl. A. Plattner and H. Wild, *ibid.,* 1946, **29**, 1611; Margrit Kobelt, P. Barman, V. Prelog and L. Ružička, *ibid.,* 1949, **32**, 356.
9. L. Ružička, Pl. A. Plattner and H. Wild, *ibid.,* 1945, **28**, 613; V. Prelog and Margrit Kobelt, *ibid.,* 1949, **32**, 1187.
10. L. Ružička, Margrit Kobelt, O. Häfliger and V. Prelog, *ibid.,* p. 544.
11. V. Prelog and O. Häfliger, *ibid.,* p. 2088.
12. Cf. K. S. Pitzer, *J. Chem. Physics,* 1940, **8**, 811.
13. H. Kuhn, *ibid.,* 1947, **15**, 843.
14. *Science,* 1945, **101**, 672.
15. E. C. Du Feu, F. J. McQuillin and R. Robinson, *J.,* 1937, 53; cf. A. L. Wilds and C. H. Shunk, *J. Amer. Chem. Soc.,* 1943, **65**, 469.
16. V. Prelog, L. Ružička, P. Barman and L. Frenkiel, *Helv. Chim. Acta,* 1948, **31**, 92.
17. V. Prelog, P. Barman and M. Zimmermann, *ibid.,* 1949, **32**, 1284.
18. *Coll. Czech. Chem. Comm.,* 1947, **12**, 93; O. Wichterle and M. Hudlický, *ibid.,* p. 101; M. Hudlický, *ibid.,* 1948, **13**, 206; O. Wichterle, J. Procházka and J. Hofman, *ibid.,* p. 300.
19. Cf. E. C. S. Jones and J. Kenner, *J.,* 1931, 1849.
20. V. Prelog and K. Wiesner, *Helv. Chim. Acta,* 1947, **30**, 1465.
21. V. Prelog, K. Wiesner, W. Ingold and O. Häfliger, *ibid.,* 1948, **31**, 1325.
22. V. Prelog and K. Wiesner, *ibid.,* p. 870.
23. V. Prelog, O. Häfliger and K. Wiesner, *ibid.,* p. 877.

NOTES

[a] The term 'constellation' denotes the distribution of atoms in space without consideration of the different forms which are possible by free rotation about simple bonds (cf. Freudenberg, *Stereochemie,* Leipzig u. Wien, 1933, pp. 535, 825).

[b] In the model shown by Fig. 9 we have 6 times the unfavourable constellation *c* and twice the favourable constellation *b*. If the oxygen atom is outside the ring, as in the model shown by Fig. 10, the unfavourable constellation *c* occurs 8 times.

[c] Dr G. B. B. M. Sutherland, Cambridge, who independently investigated the infra-red spectra of some cyclanones observed in *cyclo*octanone changes in the frequences of the C=O band at about 5·7 μ as well as of the C—H band at about 1·75 μ, which indicate an intramolecular hydrogen bond.

ASYMMETRIC METHYL GROUPS, AND THE MECHANISM
OF MALATE SYNTHASE

[from *Nature (London)* **221**, 12–13 (1969)]

Many enzyme reactions are known in which a methyl group ($-CH_3$) is converted to a methylene group ($-CH_2-$). If the steric course of such reactions is determinate, there must be a stereochemical relationship between the hydrogen atom which becomes displaced and the new group which displaces it. With the necessary changes this is also true for those enzymatic reactions in which methylene becomes methyl: discovery of these relationships throws new light on the enzymatic mechanisms.

One way to study the stereochemistry of such reactions is to use substrates, or generate products, having methyl groups made asymmetric by labelling with both deuterium and tritium. When such methyl groups are transformed by a stereospecific reaction into methylene groups, the stereochemistry can be traced if (a) the chirality[a] of the methyl group is known, (b) the chirality of the derived methylene group can be determined, and (c) there is appreciable discrimination between isotopes in the removal of hydrogen. Similarly, when a substrate containing a methylene group made asymmetric by labelling with one hydrogen isotope is transformed enzymatically into a product containing a methyl group, this group will be chiral if the enzymatic reaction is stereospecific and if the added hydrogen atom is labelled with the alternate hydrogen isotope. The steric course of the reaction can then be found if (d) the chirality of the derived methyl group can be determined, and (e) the chirality of the methylene group is known.

The problem of recognizing chirality is complicated by the impossibility, in practice, of using tritum undiluted by normal hydrogen. This seems to rule out optical rotation as a means of measuring chirality: most methyl groups in an asymmetrically labelled specimen will consist of $-CH_2{}^2H$, and this is not asymmetric. The difficulty is overcome by (i) ensuring that practically every methyl group containing tritium also contains deuterium, and (ii) relating the determination of chirality to measurement of radioactivity, which ensures that only molecules containing tritium are inspected. The isotopic discrimination required by condition (c) is then between hydrogen and deuterium.

Malate synthase (EC.4.1.3.2) produces S-malate (I) by reaction between acetyl-coenzyme A and glyoxylate. No exchange of carbon-bound hydrogen between acetyl-coenzyme A and the aqueous medium occurs during the enzymatic reaction.[1] The intramolecular tritium isotope effect is approximately 2·7, as measured by comparing the specific radioactivities of ^{3}H-acetyl-coenzyme A and of S-malate produced from it. The intermolecular deuterium isotope effect is 1·4 as measured by comparing maximum reaction velocities with acetyl-coenzyme A and with trideuteroacetyl-coenzyme A (unpublished results of G. Biedermann). Thus it was likely that condition (c) would be satisfied by this enzyme.

Fumarase (EC.4.2.1.2) catalyses the reversible dehydration of S-malate. The hydroxyl group at C_2 and a pro-R hydrogen (H^*) at C_3 are removed: a *trans* elimination.[2,3] No other carbon-bound hydrogen in malate or fumarate (II) is exchanged during equilibrium. Thus if one starts from S-malate labelled with tritium at C_3 the product at equilibrium will consist of a mixture of malate and fumarate from which all tritium originally present at the pro-R position (H^*) has been exchanged with water of the medium. All tritium originally present at the pro-S (H^0) position will still be present in malate and fumarate, and the specific radioactivities of these two acids will be equal, neglecting the much diluted radioactivity reintroduced into malate by water of the medium adding to fumarate. Thus the percentage of tritium retained in the organic acids after equilibration with fumarase is equal to the percentage of $2S$-$3S$ [3-$^3H_1$] malate in the S-malate specimen examined, and the percentage of tritium lost from the organic acids equals the percentage of $2S$-$3R$ [3-$^3H_1$] malate.

It follows that if S-malate is synthesized stereospecifically from acetyl-coenzyme A on malate synthase, and if removal of hydrogen (rather than deuterium) from acetyl-coenzyme A is favoured, the synthesis from chiral [$^2H_1$ $^3H_1$] acetyl-coenzyme A will give unequal proportions of $2S$-$3S$ [3-$^2H_1$ $^3H_1$] malate and $2S$-$3R$ [3-$^2H_1$ $^3H_1$] malate, and the loss of carbon-bound radioactivity on equilibration of this malate with fumarate on fumarase will differ from 50 per cent. The deviation will be a function of the optical purity of the acetyl groups containing tritium, and of the isotope effect in the synthesis of malate.

We prepared R and S-[$^2H_1$ $^3H_1$] acetate of known chirality by chemical procedures. Deutero-phenylacetylene (III) was prepared from deuterium oxide and phenylethynyl magnesium bromide, and was reduced by diimide to *cis*-2-$^2H_1$ -1-phenylethylene (IV), the isotopic and stereochemical homogeneity of which (complete, within the limit of accuracy) was established by nuclear magnetic resonance measurements.[4] Epoxidation by peroxybenzoic acid gave the epoxide (V*a, b*) which was reduced by lithium borotritide to 2-[$^2H_1$ $^3H_1$]-1-phenylethanol (VI*a, b*). Because epoxidation of olefins is a *cis* addition and reduction of epoxides by metal hydrides inverts configuration at the carbon atom from which oxygen is removed, the (racemic) product is a mixture of $1R$-$2R$ [$^2H_1$ $^3H_1$]-1-phenylethanol (VI*a*) and $1S$-$2S$ [$^2H_1$ $^3H_1$]-1-phenylethanol (VI*b*). At this stage the material could be diluted for convenience with unlabelled material. The alcohol was resolved into (+)-$1R$ and (−)-$1S$ enantiomers by crystallization of the brucine phthalates[5] (the assignment of absolute configuration rests securely on chemical correlations with lactic and mandelic acids.[6] Careful oxidation of each enantiomer by chromic acid destroyed asymmetry at $C(_1)$ and gave to enantiomers of [$^2H_1$ $^3H_1$] acetophenone (VII*a, b*) which were oxidized by peroxytrifluoroacetic acid to phenyl acetates (VIII*a, b*). On mild alkaline hydrolysis these gave R and S-[$^2H_1$ $^3H_1$] acetic acids (IX*a, b*) characterized as 4-bromophenacyl esters and having specific radioactivities around 0·07 μCi/μmole. All steps between VI and IX were checked for exchange of hydrogen between unlabelled substrate and labelled medium; this was slight enough to ensure that little racemization of the methyl groups had occurred during the synthesis.

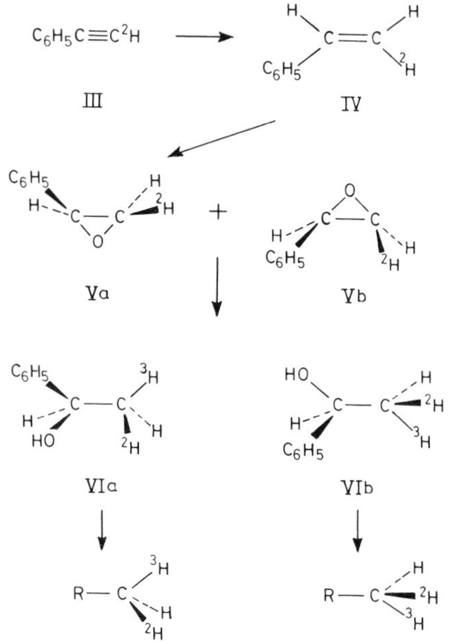

$$C_6H_5C{\equiv}C^2H$$

III

IV

Va + Vb

VIa   VIb

VIIa R = C_6H_5CO- VIIb R = C_6H_5CO-
VIIIa R = C_6H_5OCO- VIIIb R = C_6H_5OCO-
IXa   R = $HOCO-$ IXb   R = $HOCO-$

Each specimen of acetate was incubated with adenosine triphosphate, acetate kinase, phosphotransacetylase, coenzyme A, glyoxylate and malate synthase. The yield of S-malate, from acetate, was 80—90 per cent. Randomly labelled [$^3H_1$] acetate was also converted into malate. The malic acids were isolated by chromatography on 'Dowex-1' and diluted with unlabelled S-malate. Each sample was divided into two halves for determination, by two methods, of the loss of tritium on incubation with fumarase. A stoichiometrically negligible amount of ^{14}C-malate was added to one half. Aliquots were taken before, and at intervals after, addition of fumarase. The organic acids were separated from inorganic ions and, after complete removal of the aqueous medium, redissolved in water. The ratio of $^3H/^{14}C$ was determined in one set of samples; in the other, counting for ^3H alone was combined with enzyme determination (nicotinamide-adenine dinucleotide + acetyl-coenzyme A + fumarase + malate dehydrogenase + citrate synthase) of total fumarate + malate. In the conditions used, the tritium content of the organic acids became constant after about 10 min.

TABLE I

	Malate from R-[$^2H_1$$^3H_1$] acetate	Malate from S-[$^2H_1$$^3H_1$] acetate	Malate from [$^3H_1$] acetate
(a) Radioactivity of malate before fumarase (c.p.m./ μmole)	$724 \pm 1·6$	$726 \pm 2·9$	700 ± 21
Radioactivity of malate + fumarate after fumarase (c.p.m./μmole)	$480 \pm 8·2$	$223 \pm 1·7$	340 ± 6
Retention of ^3H (%)	$67·2 \pm 1·3$	$30·7 \pm 0·3$	$48·1 \pm 0·9$
(b) $^3H/^{14}C$ ratio of malate before fumarase	$1·558 \pm 0·004$	$1·560 \pm 0·018$	$1·541 \pm 0·024$
$^3H/^{14}C$ ratio of malate + fumarate after fumarase	$1·071 \pm 0·007$	$0·481 \pm 0·003$	$0·770 \pm 0·012$
Retention of ^3H (%)	$68·7 \pm 0·8$	$30·8 \pm 0·5$	$50·0 \pm 1·6$

The results summarized in Table I showed agreement between the two methods of assay. The loss of tritium from malate derived from randomly labelled [$^3H_1$] acetate was close to the expected 50 per cent, and malate synthesized from S-[$^2H_1$$^3H_1$] acetate lost more than twice as much tritium as did malate from R-[$^2H_1$$^3H_1$] acetate. The disparity corresponds to an intramolecular deuterium isotope effect of about 2·2 in malate synthesis. More than two-thirds of the tritium-containing molecules in the malate from R-[$^2H_1$$^3H_1$] acetate, and less than one-third in the malate from S-[$^2H_1$$^3H_1$] acetate, were $2S$-$3S$-[3-$^2H_1$$^3H_1$] malate. The synthesis of malate from acetate (shown below for the R-acetate) and glyoxylate therefore occurs with inversion of configuration at the methyl group.

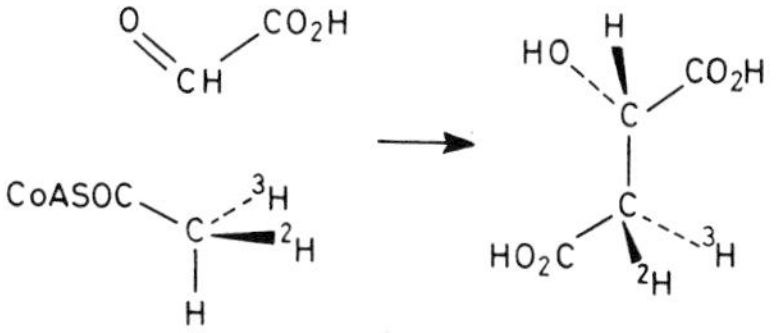

A fairly simple and decisive method is thus made available for determining the chirality of doubly labelled methyl groups, and hence for studying enzymatic reactions in which these are generated.

We thank Dr C. Donninger for useful discussion, G. Biedermann for gifts of malate synthase, and Verena Buschmeier and Keith Perrin for experimental assistance.

Shell Research, Limited,   J. W. CORNFORTH
Milstead Laboratory,   J. W. REDMOND
Sittingbourne, Kent.

Institut für Biochemie,   H. EGGERER
der Universität München,   W. BUCKEL
West Germany.   CHRISTINE GUTSCHOW

REFERENCES

1. Eggerer, H., and Klette, A., *Europ. J. Biochem.,* **1**, 447 (1967).
2. Gawron, O., Glaid, A. J., and Fondy, T. P., *J. Amer. Chem. Soc.,* **83**, 3634 (1961).
3. Anet, F. A. L., *J. Amer. Chem. Soc.,* **82**, 994 (1960).
4. Yoshino, T., Manabe, Y., and Kikuchi, Y., *J. Amer. Chem. Soc.,* **86**, 4670 (1964).
5. Downer, E., and Kenyon, J., *J. Chem. Soc.,* 1156 (1939).
6. For review of evidence see *Progress in Stereochemistry* (edit. by Klyne, W.) 187 (Butterworths, London, 1954).

NOTES

[a] A molecule is chiral, or has chirality, if its realized image in a plane mirror cannot be brought into coincidence with itself.

2

Early geometrical concepts of matter

The word stereochemistry (from the Greek word *stereos*, solid) was first used in 1890 by Victor Meyer and refers to that branch of chemistry concerned with a description of the relative positions in space of the atoms in a molecule. The word is derived from an 1875 booklet of J. H. van't Hoff entitled *La Chimie dans l'Espace* ('Chemistry in Space'). The basic concepts of stereochemistry proposed by Van't Hoff (and J. A. Le Bel) in 1874 were gradually incorporated into the theoretical structure of organic and inorganic chemistry. At the same time, it also was thought of as a separate branch of chemistry. In the earlier phases of development, chemists were interested in the 'statical' aspects of stereochemistry, that is, the investigators were concerned with the phenomenon of stereo-isomerism. By the end of the first half of the 20th century, the investigations had turned to the more 'dynamical' aspects of stereochemistry which sought to understand the relationship between the three-dimensional structure of molecules and their chemical reactivity.

The ideas of classical stereochemistry arose from difficulties encountered in the application of the valency theory to the writing of chemical formulas and to the observation of a new type of isomerism known as 'optical isomerism'. For most chemists, a discussion of the history of stereochemistry begins in the middle of the 19th century, and this book will be largely concerned with developments in this period. This does not mean that there were no scientists interested in the shape of molecules prior to the 1850s, but only that the experimental and philosophical basis for this interest was, for the most part, of a different kind. Since an atomic-molecular theory of matter was not well established until the middle of the 19th century, it is perhaps better to talk about the pre-history of stereochemistry in terms of geometrical concepts of matter.

As an example of one of the earliest forms of speculation, we might start with Plato (428–347 B.C.) who attempted to account for the

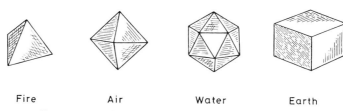

Fire   Air   Water   Earth

Figure 1. The four Platonic figures and the four elements.

interconvertibility of matter by associating each of the 'four elements' (fire, air, water, earth) with a regular geometric figure (Fig. 1). Earth was associated with the cube because of its stability; the angles on the remaining figures were considered to be related to the mobility and lightness of the elements. The interconversion of the elements was possible because the solid figures were composed of the more fundamental triangles. It is not possible at this time to discuss Plato's theory in any detail. Plato's ideas appeared in the dialogue 'Timaeus'. Although Platonic–Pythagorean theories of matter were known in the Middle Ages, the concepts did not contribute significantly to structural theories of matter until much later.

Leucippus, through his pupil Democritus (460–370 B.C.), is sometimes considered by chemists to be the forerunner of modern atomic theory. This 'atomic' theory, however, was a philosophical attempt to explain the change that was evident in the world, while at the same time provide for its permanence. Change was brought about as the result of the rearrangement and motion of the solid, enduring atoms. The atomic theory served as the basis of the materialistic philosophy of Epicurus (340–270 B.C.) as exemplified later in Lucretius' poem 'De rerum natura' (about 50 B.C.):

> Now what kinds of matter constitute
> These first-beginnings, and how different
> They are in shape, how varied in their forms.
> Many, indeed, are very much alike,
> But, as a general rule, they tend to differ . . .
>   Hard things,
> Tight-knit, must have more barbs and hooks to hold them,
> Must be more interwoven, like thorny branches
> In a hedgerow; in this class of things
> We find, say, adamant, flint, iron, bronze
> That shrieks in protest if you try to force
> The stout oak door against the holding bars.
> Fluid substances, though, must be composed
> of smooth and rounded particles . . .

This much explained, let me go on to state
A corollary truth that follows from it,
Namely, that atoms have a finite number
Of differing shapes . . .
                              . . . let's suppose one chunk of matter
Consists of three smaller parts, or maybe
Add a few more; when you take all these parts
Placing them high and low, or left and right,
[then] you have learned, by shifting them around,
Each possible arrangement, pattern, shape.

Although the atomic theory was known to some extent in the Middle Ages, active interest was rekindled only in the 15th century. A complete translation of Lucretius' poem was recovered in 1417. It was not until the early part of the 17th century, however, that atomism was considered seriously as a basis for a theory of matter. The revival of Epicurian atomism in the middle of the 17th century was due in large part to the writings of Pierre Gassendi (1592–1655) who attributed the varying properties of matter to the differing sizes and shapes of the constituent atoms.

A particulate view of matter was also suggested by René Descartes (1596–1650) in his *Discourse on Method*. The changes observed in matter were considered to be a consequence of the motion of the particles and not so much related to their shape or size. The English natural philosopher Robert Boyle (1627–1691), on the other hand, considered that the properties of matter derived from the degree of aggregation and shape of the particles formed from the more fundamental simple 'bodies' or 'elements'. By 'elements', Boyle meant

. . . certain primitive and simple, or perfectly unmingled bodies; which not being made of any other bodies, or of one another, are the ingredients of which all these perfectly mixed bodies are immediately compounded, and into which they are ultimately resolved . . . To be short, as the difference of bodies may depend merely upon that of the schemes whereinto their common matter is put; so that the seeds of things, the fire and other agents . . . partly by altering the shape or bigness of the constituent corpuscles of a body, partly by driving away some of them, partly by blending others with them, and partly by some manner of converting them, may give the whole portion of matter a new texture of its minute parts, and thereby make it deserve a new and distinct name.

Although it is possible to find in Boyle's writings what corresponds to the definition of an element, and perhaps even a suggestion that the properties of substances should be related to the shape of the corpuscles, the statement quoted above better serves to illustrate the essentially speculative nature of the 'corpuscular' philosophy.

In the 'Queries' that Isaac Newton (1642-1727) appended to later editions (1717) of his *Opticks*, are found his speculations about the ultimate structure of matter, which might strike the reader as not unsimilar to Boyle's views:

> All things being considered, it seems probable to me, that God in the Beginning form'd Matter in solid, massy, hard, impenetrable, moveable Particles, of such Sizes and Figures, and with such other Properties, and in such Proportion to Space, as most conduced to the End for which he form'd them; and these primitive Particles being Solids, are incomparably harder than any porous Bodies compounded of them . . .

The atomic theory of John Dalton (1766—1844) might be considered as an outgrowth of the corpuscular belief of his contemporaries. In fact, Dalton often quoted Newton's statement in support of his views. Dalton was interested in a more satisfactory explanation of physical properties of matter. By 1803, he had concluded that the solubility behaviors of gases in liquids depended on the properties of the constituent particles: '. . . I am nearly persuaded that the circumstances depend upon the weight and number of the ultimate particles of the several gases . . .' The extension of these atomic ideas to the composition of matter was developed in the next few years. The definite (and multiple) proportions of the constituent elements found in many compounds could be accounted for by assigning a definite mass to the atoms of each element.

Although chemists acknowledged that the atomic theory provided a rationalization for the observation of definite and multiple proportions in compounds, complete acceptance of Daltonian atomism was delayed until the middle of the 19th century because of experimental difficulties and what were thought to be the more speculative features of the atomic theory. The procedure used by Dalton to determine molecular formulas led to a lack of agreement as to the atomic weights of elements. The reader of the chemical literature in the first half of the 19th century will, therefore, encounter many formulas in which the number of elements is halved, doubled, or some multiple of that presently accepted. The atomic weight 'problem' generated a certain scepticism about atomism with the result that often the theory was referred to as the 'Law of Chemical Combination'. To many, then, the formulas indicated only the 'combining' or 'equivalent' weights of the constituent elements. It was clear that the formula of a compound would provide little information about the properties of a substance. William Hyde Wollaston, an early supporter of

Dalton's theory, commented in 1808 that the theory could be considered of limited value if it could only explain the phenomenon of definite proportions:

> . . . I am inclined to think, that when our views are sufficiently extended, to enable us to reason with precision concerning the proportions of elementary atoms, we shall find the arithmetical relation alone will not be sufficient to explain their mutual action, and that we shall be obliged to acquire a geometrical conception of their relative arrangement in all three dimensions of solid extension.

He proposed that there would be a 'natural' arrangement of compounds containing only two elements (A and B): for AB_2, linear; AB_3, trigonal; and AB_4, tetrahedral. Since the surrounding particles might disturb these arrangements, he concluded that '. . . It is perhaps too much to hope that the geometrical arrangement of primary particles will ever be known'.

Five years later Wollaston extended his speculations on the arrangements of atoms in space by considering how these atoms might be used to form regular crystallographic forms. In 1690, Christian Huygens had also demonstrated how spherical atoms could be used to construct crystalline forms. Wollaston's proposal, however, provided an alternative to the theory proposed in 1784 by the Abbé René Just Haüy (1743–1822). Briefly summarized, Haüy proposed that the crystal shape was macroscopically equivalent to the shape of the 'constituent molecules' composing the crystal. Originally he reduced the shapes of these constituent molecules to six 'primitive forms', whose shape could be discovered by the mechanical division of the crystal.

Wollaston's support for, and geometrical extension of, the atomic theory did not continue. Within a few years he had abandoned the use of 'atomic' weights in favor of 'equivalent' weights to calculate the composition of compounds. He even developed a kind of chemical slide rule to facilitate the calculations.

Dalton's geometrical extension of the atomic theory produced atom arrangements that suggested particular structures for molecules. This is particularly apparent in the two-dimensional illustrations found in Dalton's notebooks, lecture diagrams and other publications (Fig. 2). Generally, the arrangements of the atoms (represented by circles) are those that place atoms of the same element as far apart as possible to minimize what Dalton felt was their 'natural' repulsion (see for example the formulas of nitrous oxide, 'sulfuretted hydrogen' and carbonic acid). It is sometimes difficult to understand why a particular arrangement is chosen for more complex molecules. In the formula of sugar (then

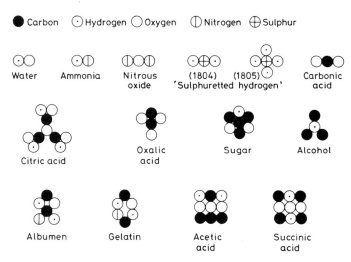

Figure 2. Examples of two-dimensional formulas used by John Dalton.

thought to be $C_4 HO_2$) the arrangement accounts for the formation of alcohol ($C_3 H$) and carbonic acid (CO_2) on fermentation.

Today, the phenomenon of isomerism is easily accounted for in terms of the different arrangements of atoms in a molecular structure. What is obvious today was not so obvious in the first half of the 19th century. First, there were few experimental reports that could be cited that might suggest the necessity of the isomerism concept. Secondly, it was not always clear whether the two substances being compared had the same molecular formula or empirical formula. For example, in 1825 Michael Faraday noted that the isobutylene isolated from compressed oil-gas had the same empirical formula as 'olefiant' gas (ethylene) but different physical properties. The synthesis of urea from ammonium cyanate in 1828 by Wöhler was of greater significance as an example of isomerism than it was toward the destruction of the concept of 'vital' force. Although it might seem to us that an explanation of the existence of isomers might be found in considering structural differences in the molecules, no progress could be made in this direction until the development of valency theory. Isomerism was not recognized as a general phenomenon until 1832 when a definition was proposed by the influential chemist, Jöns Jacob Berzelius (1779–1848):

In physical chemistry it has long been considered an axiom that substances composed of the same constituents and of the same relative quantities of these must of necessity also have the same chemical properties. Experiments of Faraday seemed to show that an exception to this axiom might occur if two bodies have the same composition, but differ in that, although the relative proportions between their elements is the same, the one contains twice as many simple atoms as does the other . . . More recent experiments have further shown that the absolute as well as the relative number of atoms can be equal, and yet that their combinations may take place in a manner so different that the properties of bodies having absolutely like compositions may become unlike . . . If further investigations should confirm this view, an important step would have been taken in the advancement of our theoretical knowledge concerning the compositions of substances. But since it is requisite that we should be able to express our conceptions by definite and appropriately chosen terms, I have proposed to call substances of the same compositions and of different properties 'isomeric', from the Greek word for 'composed of parts'.

Chemists' reluctance to use the atomic symbols of Dalton made it more difficult for them to explain how compounds having the same composition might exhibit different properties. Even in Dalton's case, however, we encounter only one example of his use of different arrangements of atoms to explain the existence of two compounds (gelatin and albumin in Fig. 2) thought to have the same composition, $C_2 H_2 ON$. No record has survived concerning Dalton's reasons for the two formulas shown.

Thomas Thomson's textbook *A System of Chemistry*, published in 1807, contains two structures for acetic acid and succinic acid, which were also believed to have the same formula, $C_4 H_2 O_3$ (Fig. 2). Thomson commented briefly on the atomic arrangements he had chosen: 'And it is obvious enough that these 9 atoms might arrange themselves in a great variety of binary compounds, and the way in which these binary compounds unite may, and doubtless does, produce a considerable effect upon the nature of the compound formed.' The two arrangements of the acids '. . . undoubtedly would produce a great change in the nature of the compound'.

Dalton did not restrict his atomic arrangements to two dimensions. In his lectures, he often used atomic models. The wooden atom spheres (Fig. 3) contained metal pins to which could be attached other atom spheres in a symmetrical fashion. The geometries proposed by Dalton for simple two-element compounds were similar to those suggested by Wollaston: AB_2, linear; AB_3, trigonal; AB_4, tetrahedral (or square planar); AB_5, a triangular pyramid; and AB_6, octahedral. It should be pointed out that two-dimensional and three-dimensional arrangements were rather arbitrarily chosen by Dalton — more from symmetry considerations than

Figure 3. Dalton's atomic models (Science Museum, London).

from any attempt to relate the arrangement to the properties of the substance. Perhaps the phenomenon of isomerism might have been more readily explained in the second quarter of the 19th century had other chemists used the Dalton symbols. Few chemists, however, shared Dalton's interest in atomic arrangements. But this should not be surprising since few chemists completely accepted the atomic theory itself, as can be seen, for example, by the lack of consideration for some 40 years to Avogadro's demonstration (1811) of the usefulness of the atomic theory in explaining Gay-Lussac's Law of Combining Volumes. The few who did draw attention to Avogadro's ideas were also ignored. In 1833, for example, Marc-Antoine Gaudin (1804–1880) used the Daltonian symbols to illustrate the formation of water from oxygen and hydrogen according to Avogadro's proposal (Fig. 4). Later in his life, Gaudin also became interested in relating the three-dimensional arrangement of atoms in substances to their crystalline forms – an approach similar to that undertaken by Wollaston earlier. Although Gaudin's publications were

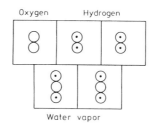

Figure 4. M.-A. Gaudin's illustration of the formation of water from hydrogen and oxygen.

known to some chemists in the first half of the 19th century, Gaudin's ideas contributed little to the development of stereochemistry in the 1870s. In part, this was due to Gaudin's lack of consideration for the development in organic chemistry in the 1850s and 1860s — particularly as regards the theory of valency. In 1873, Gaudin summarized the results of his publications in a book entitled *The Architecture of the World of Atoms*. However, since it did not deal with contemporary concepts in organic chemistry, Gaudin's ideas did nothing to contribute to the formulation of basic stereochemical ideas of the sort advanced by Van't Hoff and Le Bel a year later. (It is interesting to note, however, that Gaudin attributed the optical isomerism found in the tartaric acids to different three-dimensional arrangements of the constituent atoms in the two acids.)

Gaudin, along with André Marie Ampère (1775–1836) and Auguste Laurent (1807–1853), form a small group of crystallographers in the early part of the 19th century who felt that the key to the understanding of chemical phenomena was geometrical — that is, in the arrangement of the atoms in chemical substances. Although crystallography and molecular geometry did find a relationship (as we may note in the studies of Mitscherlich and Pasteur), the studies of this group seemed so highly speculative that they were of little use to most organic chemists.

3

Valency and chemical structure

In the first half of the 19th century, the chemical formula indicated only the kind and number of elements in a compound. Since no basis existed by which chemists could agree on the atomic weight of an element, organic chemists in particular had to contend with more than one formula for the same compound. Stanislao Cannizzaro's solution to the atomic weight problem advanced at the Karlsruhe conference in 1860 had an immediate impact on the subsequent development of structural organic chemistry. Julius Lothar Meyer commented on his return from the conference:

> As soon as the apparent discrepancies between Avogadro's rule and that of Dulong and Petit [a rule that relates the atomic weight to the specific heat of an element] had been removed by Cannizzaro, both were found capable of practically universal application, and so the foundation was laid for determining the valences of the elements, without which the theory of atomic linking could certainly never have been developed.

Although efforts had been directed earlier to the writing of chemical formulas that might convey more information about a substance's chemical behavior, the activity largely centered upon the classification of organic substances. Thus the 'radical' theory treated certain groups of elements, one of which was carbon, as analogs of elements that might be found to pass through a series of chemical transformations unchanged. For example, the benzoyl radical, C_7H_5O, could be located in the formulas for benzoic acid, benzoyl chloride and benzaldehyde, all of which could be interconverted by chemical reactions:

$$C_7H_5O \cdot OH \longrightarrow C_7H_5O \cdot Cl \longrightarrow C_7H_5O \cdot H$$

Benzoic acid   Benzoyl chloride   Benzaldehyde

By 1837, Justus Liebig (1803–1873) and Jean Baptiste André Dumas (1800–1880) were convinced that a unifying principle of organic chemistry had been found:

> In mineral chemistry the radicals are simple; in organic chemistry the radicals are compound; that is all the difference. The laws of combination and of reaction are otherwise the same in these two branches of chemistry.

Within a short period of time, however, the theory had been undermined by the observation that substitution of hydrogen in the compound radical could be affected without apparently changing significantly the chemical properties of the substance.

The similarity in chemical behavior was expressed by means of 'type' formulas. Dumas stated the theory in 1839: 'In organic chemistry there are certain types which remained unchanged, even when their hydrogen is replaced by an equal volume of chlorine, bromine and iodine.' Thus, acetic acid and trichloroacetic acid exhibited similar chemical behavior and could be considered to belong to the same type. The chemical behavior was presumably related to the arrangement of the atoms: A compound's '. . . chemical character is dependent primarily on the arrangement and number of its atoms and in a lesser degree on their chemical nature'. The arrangement was only partially represented in the 'type' formulas that were developed several years later. For example, amines were considered to be part of an 'ammonia type' which were illustrated by bracket formulas:

$$\left.\begin{array}{c} H \\ H \\ H \end{array}\right\}N \qquad \left.\begin{array}{c} H \\ H \\ C_2H_5 \end{array}\right\}N \qquad \left.\begin{array}{c} H \\ C_2H_5 \\ C_2H_5 \end{array}\right\}N \qquad \left.\begin{array}{c} C_2H_5 \\ C_2H_5 \\ C_2H_5 \end{array}\right\}N$$

Ammonia Ethylamine Diethylamine Triethylamine

More explicit structural ideas awaited the development of valency theory in the late 1850s. The idea of valency had been introduced in 1852 by Edward Frankland (1825–1899):

> *The combining-power of the attracting element . . . is always satisfied by the same number of atoms.* It was probably a glimpse of the operation of this law amongst the more complex organic group which led Laurent and Dumas to the enunciation of the theory of types; and had not those distinguished chemists extended their views beyond the point to which they were well supported by then existing facts, had they not assumed that the properties of an organic compound are dependent upon the

Plate 5. Friedrich August Kekulé von Stradonitz (1829–1896).

position and not upon the *nature* of its single atom, that theory would undoubtedly have contributed to the development of the science to a still greater extent than it has already done.

The more explicit statement of valency was proposed in 1858 independently by Friedrich August Kekulé (1829–1896) (Plate 5) and Archibald Scott Couper (1831–1892). The great diversity of carbon-containing compounds could be accounted for by this theory. As Kekulé wrote:

> If only the simplest compounds of carbon are considered . . . it is striking that the amount of carbon which the chemist has known as the least possible, the atom, always combines with four atoms of a monoatomic, or two atoms of a diatomic element; that generally, the sum of the chemical units of the elements which are bound to one atom of carbon is equal to four. This leads to the view that carbon is tetraatomic (tetrabasic). . . . it must be assumed that at least part of the atoms are held just by the affinity of the carbon, and that the carbon atoms themselves are joined together, so that naturally a part of the affinity of one for the other will bind an equally great

part of the affinity of the other. The simplest, and therefore the most obvious, cases of such linking together of two carbon atoms is this, that one affinity unit of one atom is bound to one of the other. Of the 2 × 4 affinity units of the two carbon atoms, two are thus used to hold both atoms together; there still remain six extra which can be found by the atoms of the other elements.

Kekulé later recounted how these ideas came to him during a visit to London in 1856:

> One fine summer evening I was returning by the last omnibus, 'outside', as usual, through the deserted streets of the metropolis, which are at other times so full of life. I fell into a reverie (*Träumerei*), and lo, the atoms were gambolling before my eyes! Whenever, hitherto, these diminutive beings had appeared to me, they had always been in motion; but up to that time I had never been able to discern the nature of their motion. Now, however, I saw how, frequently, two smaller atoms united to form a pair; how a larger one embraced two smaller ones; how still larger ones kept hold of three or even four of the smaller; whilst the whole kept whirling in a giddy dance. I saw how the larger ones formed a chain, dragging the smaller ones after them, but only at the ends of the chain. I saw what our Past Master, Kopp, my highly honoured teacher and friend, has depicted with such charm in his 'Molekularwelt'; but I saw it long before him. The cry of the conductor: 'Clapham Road', awakened me from my dreaming; but I spent a part of the night in putting on paper at least sketches of these dream forms. This was the origin of the *Strukturtheorie*.

Kekulé's description of the different atom sizes and the manner in which they linked suggests that the 'sketches of these dream forms' may have been the source of the graphic 'roll' formulas that appeared in his textbook a few years later (Fig. 5).

Couper's papers also developed the valency concept to explain the linking of carbon, but due to a delay in the publication of Couper's paper and a priority claim by Kekulé, Couper's ideas remained relatively unknown to most chemists. It seems curious that Couper's papers were not better known, since to the modern eye, the graphic formulas (Fig. 9) used by Couper convey more structural information than the papers published by Kekulé. There is some evidence that Couper did have a profound influence upon A. M. Butlerov, who was responsible for the introduction of the theory of 'chemical structure'.

The term 'valency' did not come into general use until the late 1860s. Some of the other terms that were used having the same meaning were 'combining power', 'degree of affinity', 'basicity', 'equivalency' and 'atomicity'. In 1866, Frankland was the first to speak of a 'bond' between elements, although he was careful to point out that he was not attributing any physical reality to the picture suggested by the use of the term:

> By the term bond, I intend merely to give a more concrete expression to what has received various names from different chemists, such as an atomicity, an atomic power, and an equivalence . . . It is scarcely necessary to remark that by this term I do not intend to convey the idea of any material connection between the elements of a compound, the bonds actually holding the atoms of a chemical compound being, as regards their nature, much more like those which connect the members of our solar system.

In 1870 he revised the latter phrase to say '. . . the bonds which actually hold the constituents together being, as regards their nature, entirely unknown'.

We can see in this statement the reluctance with which organic chemists were willing to make any claims concerning the actual structure of molecular species — since it was not believed that a knowledge of the physical arrangement of atoms could be gained from only chemical studies.

The term 'chemical structure' was introduced by Alexander Mikhailovich Butlerov (1828–1886) in 1861:

> It seems to be quite natural that chemistry, which deals only with substances during the course of their transformation, is unable to make any statements about their mechanical structure, at least as long as there is no assistance from physical investigation . . . To be sure we don't know what relationship exists between the chemical influence and the mutual mechanical arrangement of the atoms in the interior of the molecule. We do not even know whether in a complex molecule two atoms which directly influence each other chemically are really situated immediately side by side. But even disregarding completely the conception of the *physical* atoms we cannot deny that the chemical properties of a substance are especially conditioned by the chemical cohesion of the elements which compose them. Let's start from the assumption that each *chemical* atom contains only a definite and limited amount of that chemical force (affinity) with which it takes part in the formation of a com-

pound. Then I would like to call *chemical* structure the chemical cohesion or the manner of the mutual binding of the atoms inside a compound substance. The well-known rule which states that the nature of a compound molecule is determined by the nature, quantity, and the arrangement of its components could then meanwhile be modified in the following way. The chemical nature of a compound molecule depends on the nature and quantity of its elementary constituents and on its chemical structure.

Butlerov's theory focused on the dynamical relationship between the state of the reacting molecule and the product. In contrast, Kekulé and Couper considered that the analysis of the components of the reactant molecule should lead to an understanding of the molecule as a whole. The latter approach led more naturally to the use of two-dimensional pictures (*graphic formulas*) as a way of visualizing the valency concept.

Two kinds of graphical pictures appeared in the 1860s. The first were formulas included as footnotes in the first volume of Kekulé's textbook of organic chemistry published in 1861 (Fig. 5) and were thought by some to resemble 'bread rolls'. In these 'roll' formulas, the 'combining power' (that is, valency) of an element was indicated by the length of the roll. Thus the carbon roll is four times as long as hydrogen, nitrogen three times, and oxygen two times as long. The 'saturation' of valencies takes place only vertically, rather than horizontally. Multiple bonds (as is illustrated in the formula for acetic acid) are produced by the contact between the valency units in several places. These graphic pictures were apparently based on

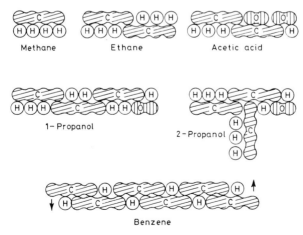

Figure 5. Graphical formulas used by A. Kekulé in 1861.

wooden models used by Kekulé in his lectures in Ghent in 1857. Although these formulas were only intended to indicate the saturation of valencies, some structural ideas are suggested. For example, Kekulé states that the two unsatisfied valencies of benzene could be saturated by forming a ring (the unsatisfied valencies in benzene are symbolized by arrows). To explain the existence of the structural isomers, 1-propanol and 2-propanol, Kekulé had to violate his rule concerning vertical saturation of valencies to allow for the placing of a methyl group branching off the main carbon chain.

The benzene formula was suggested in 1865. A triangular structure (as would be easily formed by joining the two ends of the benzene model) was considered but rejected since an excessive number of mono- and disubstituted derivatives would be possible. The hexagonal benzene structure was introduced in 1866. Kekulé later maintained that another dream (in 1863) was responsible for the birth of his benzene structure:

> During my time in Ghent I occupied smart bachelor quarters in the main street. My study however lay against a side lane and was deprived of light during the day, but this was of little disadvantage to a chemist whose daylight hours were spent in the laboratory. There I sat writing my textbook, but to little avail: my thoughts were upon other things. I turned my chair to the fire and dozed. Again the atoms gambolled before my eyes, but this time the small groups kept demurely in the background. My mind's eye, sharpened by many previous experiences, distinguished larger structures of diverse forms: long series, closely joined together: all in motion, turning and twisting like serpents. But see, what was that? One serpent had seized its own tail, and this image whirled defiantly before my eyes. As by a lightning flash I awoke: and again spent the rest of the night working out the consequences of this idea.
>
> Let us learn to dream, gentlemen, and then we may find the truth . . . but let us beware of making our dreams public before they have been approved by the waking mind.

Although dreams and imagination served as the stimulus for Kekulé's structural ideas, Kekulé made it clear that one should not confuse the models with reality:

> . . . Some chemists are still of the view that from a study of chemical metamorphoses one can derive the constitution of compounds with certainty and can express in a chemical formula this position of the atoms. That this last is not possible does not need special proof; it is self-evident that one cannot

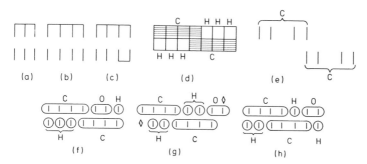

Figure 6. Some valence structures (after Kekulé) used in the 1860s: George Cary Foster (1863): (a) NH_3, (b) CH_4, (c) CH_2O; Charles Wurtz (1869): (d) CH_3CH_3; J. Wilbrand (1865): (e) CH_3CH_3; Alfred Naquet (1868): (f) aldehyde (ethanal), H_3CCHO, (g) oxide of ethylene, $\cdot CH_2CH_2O\cdot$, (h) acetylic alcohol (vinyl alcohol), $H_2C=CHOH$.

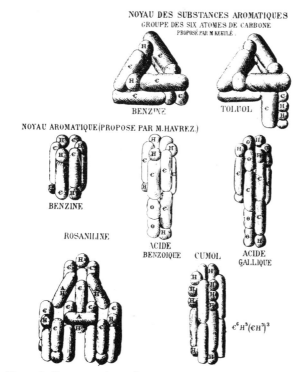

Figure 7. Benzene valence formulas proposed by Havrez in 1866.

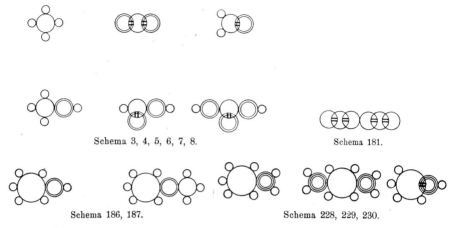

Schema 3, 4, 5, 6, 7, 8.   Schema 181.

Schema 186, 187.   Schema 228, 229, 230.

Figure 8. Loschmidt's formulas: 3, methane; 4, carbon dioxide; 5, formaldehyde; 6, methyl alcohol; 7, formic acid; 8, carbonic acid; 181, benzene nucleus (trial version); 186, phenol; 187, methoxybenzene; 228, aniline; 229, 1,4-diaminobenzene; 230, a hypothetical benzene imide.

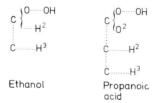

Figure 9. A. S. Couper's formulas of ethanol and propanoic acid.

Figure 10. A. Crum Brown's formulas of ethane and vinyl chloride.

show the position of atoms in space, even if one had investigated this, on the plane of the paper by putting letters together; for this one would need at least a perspective drawing or a model. But it is likewise clear that one cannot determine the position of atoms in a specific compound by a study of meta-

morphoses, because the way in which the atoms leave a changing and decomposing compound cannot indicate how they are
arranged in the existing and unaltered compound. Certainly it
must be considered a problem for research workers to discover
the constitution of the materials, and thus, if you will, the
position of the atoms, but this can certainly not be accomplished by the study of chemical changes, but only by comparative
studies of the physical properties of the unchanged compounds.

This view was shared by many chemists of the period and accounts in
part for the difficulties encountered by Van't Hoff in the acceptance of
his stereochemical ideas in the 1870s. Although Kekulé stated that the
models were used by a number of chemists, little information is available
to confirm this. The graphical pictures were used in the 1860s by only a
few chemists (Fig. 6). Although these were intended only as valency
formulas, some structural considerations are suggested in some of the
formulas — in particular in the formulas written by Alfred Naquet for
three isomeric compounds.

The three-dimensional figures used by P. Havrez in 1866 suggest the use
of models (Fig. 7). Note the alternative to the Kekulé structure proposed
by Havrez. In 1861 the chemist Josef Loschmidt (1821–1895) published
a pamphlet, *Chemical Studies*, which included illustrations of some 368
chemical formulas. The atoms of different elements were represented by
circles, with the valency saturation indicated by the touching of the
circles (Fig. 8). Multiple bonds were indicated by intersecting circles.
Some of Loschmidt's formulas of benzene anticipated Kekulé's ring
structure. The pamphlet was privately printed and apparently did not
receive a wide enough distribution to contribute significantly to the
development of structural chemistry. However, Kekulé's reference in a
letter to Erlenmeyer in 1862 to 'Loschmidt's confusion-formulae' does
indicate his awareness of Loschmidt's ideas.

The formulas (Fig. 9) used by Couper in 1858 suggest a more precise
arrangement of the atoms than is found in the Kekulé formulas. The
number of oxygens found in Couper's formulas are double what they
should be since Couper used an atomic weight of oxygen = 8. A few years
later, in 1864, Alexander Crum Brown (1838–1922) at Edinburgh University (where Couper had also been for a short time) published a paper
which included what he called 'constitutional formulae' to represent the
chemical properties of compounds (Fig. 10). An element's valency (the
term 'equivalent' was used by Brown) was illustrated by dashes; the
'saturation' of valency was illustrated by the orientation of the dashed
lines toward each other. For unsaturated compounds, such as vinyl
chloride, the valency lines had to be 'bent'. (In his doctoral thesis earlier,
the atoms were connected by dotted lines, in a manner similar to that

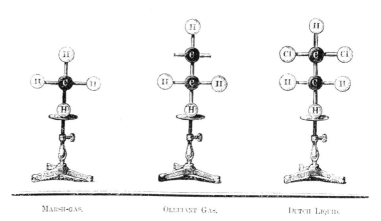

Figure 11. Croquet ball models used by A. Hofmann in 1865 (models of methane, ethylene and ethylene 'dichloride').

found in Couper's formula.) When the 1864 paper was reprinted a year later in the *Journal of the Chemical Society*, the atoms were connected by a solid line. More structural information was suggested by these formulas than those employed by Kekulé or Couper. Nevertheless, Crum Brown was insistent that the formulas illustrated only the 'chemical' positions of the atoms, and not their 'physical' or actual positions.

These formulas also suggest a ball-and-stick kind of model, and it was not long before such models were used. In 1865, August Hofmann (1818–1892) gave a lengthy talk at the Royal Institution in which the formulas were illustrated with models made from croquet balls (Fig. 11). The number of pegs from the variously painted croquet balls was equal to the 'combining power', or valency, of the elements. Carbon, for example, had four pegs which were at 90° from each other in the same plane. An early structural problem that arose from the use of these models can be seen in the formula given for ethylene ('olefiant gas'). Since the two unsatisfied valencies were placed on the same carbon, an incorrect structure is given to the chlorination product (the 'Dutch liquid': 1,2-dichloroethane).

In 1867 an editorial note appeared in the magazine, *The Laboratory*, which described 'glyptic formulae' models available from Mr Blakeman in Gray's Inn Road, London. The models (Fig. 12) consisted of variously colored wooden spheres that could be connected by brass rods or rubber tubing:

> The figures that may be formed by the combination of these coloured balls are very striking and are more likely to rivet the attention of students than chalk symbols on a blackboard.

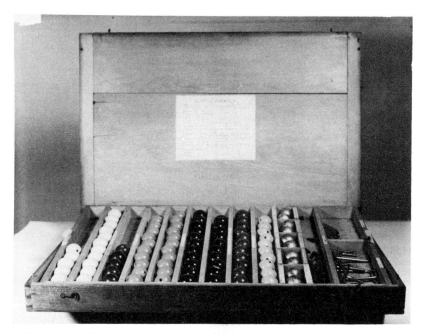

Figure 12. 'Glyptic formulae' models (Museum of the History of Science, Oxford).

Whether they are calculated to induce erroneous conceptions is a question about which much might be said.

The use of such models and formulas was viewed with some concern by many chemists in the 1860s and touched off a controversy concerning the validity of the atomic theory itself. Nevertheless, graphic formulas and models continued to be used and undoubtedly contributed significantly to the development of structural organic chemistry.

In 1867, for example, James Dewar (1842–1923) prepared a set of valency models consisting of brass bars which, when two of them were joined together at the center, represented the tetravalent carbon atom (structure 1 in Fig. 13). Dewar then showed how these models could be used to construct seven possible structures of benzene. The so-called 'Dewar benzene', found in the lower right-hand corner of the figure, is only one of these and does not represent an alternative proposal to the Kekulé formula.

After Dewar had spent a short time in Kekulé's laboratory in Ghent, he returned in 1867 bringing with him some new models that Kekulé had prepared. In these new Kekulé models, the carbon atom consisted of a

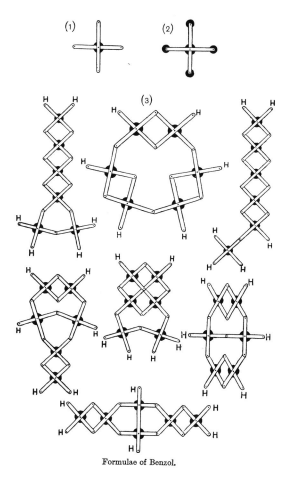

Figure 13. Valency models used by James Dewar in 1867.

wooden sphere from which emerged four tetrahedrally oriented wires. Kekulé had prepared these new models because he was dissatisfied with the manner in which the Crum Brown formulas portrayed unsaturated compounds. In the illustrations of the triple bond, the valency lines must be bent — which implies a bending of the valency forces. With the new models, Kekulé could produce a triple bond by the touching of three of the four tetrahedrally oriented valency wires. (This would correspond to the sharing of faces of the Van't Hoff tetrahedra.) The touching wires were joined by either a slit-tube fastener (see the Paternò models in Fig. 16) or by a small ring looped through holes drilled at the end of the

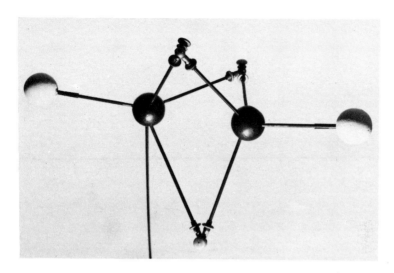

Figure 14. **Kekulé–Von Baeyer** model of acetylene, HC≡CH (the vertical wire is used to support the model).

valency wires. In about 1885, Adolf Baeyer modified these models by replacing the connecting ring with an adjustable joint (Fig. 14). Descriptions of what are called Kekulé–von Baeyer models appear in the catalogs of scientific supply companies well into the 1930s. The model shown in Fig. 14 was constructed with models used in the Department of Chemistry at Eastern Michigan University in the 1920s or 1930s. It is probable that these tetrahedral models were used by other chemists before 1874, the year in which Van't Hoff introduced his theory of the tetrahedral carbon. There is no evidence that Van't Hoff or Le Bel were aware of these models. Since it can be shown that a number of chemists were using tetrahedral carbon models before 1874, it is perhaps surprising that Van't Hoff's proposal of the tetrahedral carbon atom was considered such a revolutionary idea by chemists.

Wilhelm Koerner (1839–1925) was a student in Kekulé's laboratory in Ghent in 1866–67. When Koerner returned to Palermo, Italy, he undertook a number of studies concerned with the structure and reactions of benzene derivatives. This research led to the development of his well-known method for experimentally determining the relative positions of the *ortho, meta* and *para* substituents in disubstituted benzenes. These studies also indicated that substituents in the *ortho* and *para* positions had a greater effect in certain kinds of aromatic substitution reactions than did a *meta* substituent. To explain the *ortho/para* substituent effect, he proposed in 1869 a 'space-filling' structure of benzene (Fig. 15). In this

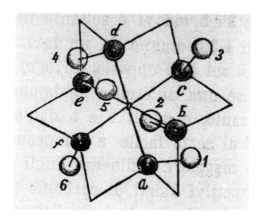

Figure 15. W. Koerner's benzene model (1869).

figure, the carbon atoms (dark spheres designated by letters) are roughly in what would now be described as a 'chair' conformation. Three of the hydrogen atoms (numbered spheres) are above, and three below the average plane made by the carbon atoms. Note also that the *ortho* and *para* carbon atoms are directly bound to each other. The construction of Koerner's benzene with Kekulé–von Baeyer tetrahedral models confirms the structural features discussed above.

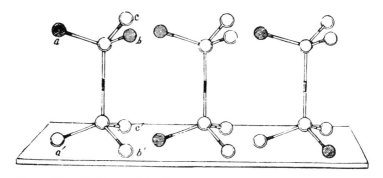

Figure 16. Models used by E. Paternò in 1869 to illustrate the three 'isomers' of dibromoethane, $C_2H_4Br_2$.

In an article in the same year and the same journal, a colleague of Koerner, Emanuel Paternò (1847–1935), included illustrations (Fig. 16) of tetrahedral carbon models to explain what he thought were the structures of three isomers of dibromoethane:

. . . one of the fundamental principles of the theory of the constitution of organic compounds, based on the atomicity of the elements and particularly on the notion of the tetra-atomicity of carbon, is that of the identical chemical function of the four valencies of the carbon atom, which is not possible unless there exists only one methyl chloride, one methyl alcohol, etc. Now the existence of isomers for compounds of the formula C_2HCl_5 cannot be explained without renouncing the idea of the equivalence of the four affinities on the carbon atom. And this is the only example known until now that is opposed to such a generally adopted idea. As for the three $C_2H_4Br_2$ isomers, given that they really exist, they are easily explained without having to assume a difference between the four affinities of the carbon atom, as Butlerov believes, when the four valencies of the atom of this element are assumed to be arranged in the sense of the four angles of a regular tetrahedron. Then the first modification would have the two bromine atoms (or any other monovalent groups) connected to the same carbon atom; while in the other modifications each of the two bromine atoms would be bonded to a different carbon atom, with only the difference that in one of the two cases the two bromine atoms would be arranged symmetrically but not in the other. This is made clearer with the following drawing in which the bromine atoms are represented by a and b.

The two 'modifications' of 1,2-dibromoethane would only be possible if rotation about the carbon–carbon bond was somehow restricted. It should be pointed out that whether rotation about single bonds was 'free' or 'restricted' was not considered a legitimate question until the publication of Van't Hoff's stereochemical ideas in 1874. It is interesting to note, however, that Paternò did not consider the possible 'staggered' conformations. Van't Hoff stated later that he was unaware of Paternò's publication. The use of tetrahedral models by Koerner was never brought to Van't Hoff's attention.

In the German edition of his *Textbook of Organic Chemistry* published in 1868, A. M. Butlerov had noted that if the existence of three isomers of tribromo- or trichloroethane was confirmed, the present ideas of chemical structure might have to be re-examined to allow for the non-equivalence of the carbon valencies. Somewhat earlier Butlerov had, in fact, proposed such an idea to explain the existence of two hydrocarbons having the formula C_2H_6. Butlerov suggested that the unequal carbon 'affinities' (or valency forces) might be oriented toward the faces of a tetrahedron. By 1868, however, the identity of the two C_2H_6 hydrocarbons had been established. This forced Paternò to consider (in the quotation above) the tetrahedral orientation of the (equivalent) carbon valencies.

The novelty of these ideas is suggested by a preface to Paternò's paper written by Cannizzaro, who stated that he planned to undertake further experimental studies. The results of Cannizzaro's investigations were never published. While it was realized that further experimental work was needed (it was shown later that only two dibromoethane isomers existed), many chemists considered any speculation about atomic arrangements to be dangerous. Adolf Lieben warned Paternò that by '. . . plunging oneself into space in the pursuit of atoms, one would risk losing the ground under one's feet'.

By the late 1860s and early 1870s, it would seem that at least a few chemists were willing to consider a three-dimensional arrangement of atoms in molecules as a way of explaining an unexpected number of isomers. To the extent that tetrahedral carbon atom models were available and used, we might assume that they contributed to the development of stereochemical concepts before Van't Hoff and Le Bel in 1874. It is interesting to note that Victor Meyer (1848–1897) stated that he had used the Kekulé models in the 1870s in order to explain the existence of only one isomer of dichloromethane. However, most of the problems faced by organic chemists did not require a stereochemical approach. The graphic formulas in use sufficed to explain the great variety of organic compounds known. Positional isomers could be portrayed with these formulas without requiring a commitment to the spatial disposition of the atoms. The structural inadequacies of the graphic formulas were only perceived when the examples of another kind of isomerism, 'optical isomerism', became too numerous to ignore.

4

Optical activity and stereoisomerism

In the first half of the 19th century various aspects of geometrical optics such as diffraction, interference and the polarization of light were of interest to a number of scientists, especially in France. Although a relationship between the phenomenon of optical rotation and the structure of matter was assumed by the early investigators, the experimental studies that ultimately served as the key to an understanding of the relationship were those of Pasteur in the late 1840s and Wislicenus in the late 1860s.

The phenomenon of polarized light was suggested in 1690 in the studies of Christian Huygens (1629–1695) who, as a result of his study of the behavior of light through crystals of Iceland spar (calcite, a form of $CaCO_3$), concluded that the crystals imparted to the waves of light 'a certain form or disposition' which enabled them to give either a single or double image. (The double refraction of light by Iceland spar was first observed by Erasmus Bartholinus in 1669.) In 1809–10, E. L. Malus noted that the light reflected from a transparent surface such as water or glass '. . . has all the characteristics of one of the beams produced by the double refraction of a crystal of spar. . .'. Malus used the term 'polarization' to describe the character of the light. In 1821, Augustin Fresnel stated that the vibrations of light could not be longitudinal, but must be transverse, and 'that the vibrations of a polarized beam must be perpendicular to what is called its plane of polarization'.

The French physicist, Jean Baptiste Biot (1774–1862) (Plate 6), was the first to investigate the phenomenon of optical activity. In a series of studies undertaken from 1811 to 1817, he observed the rotation of the plane of polarization when polarized light was passed through a piece of quartz that had been cut perpendicular to its axis. The amount of rotation was proportional to the thickness of the quartz. (By 1818, he had also demonstrated that the rotation was '. . . inversely proportional to the squares of the lengths of their vibrations in the system of waves' (that is,

Plate 6. Jean Baptiste Biot (1774–1862) (Edgar Fahs Smith Collection).

$\alpha = k/\lambda^2$, where α = the rotation, k = a constant, and λ = the wavelength of the light).

One of the more interesting discoveries made in this period was that the optical activity was not restricted to crystalline substances, but was observed in certain liquids and solutions. The optical activity of some natural liquids (oil of turpentine, oil of lemon, and camphor dissolved in alcohol) was reported in 1815–17. The optical activity of sugar solutions was observed in 1818. (The 'inversion' of the optical rotation of an acidified sugar solution due to the hydrolysis of the sugar was not observed until 1832.)

Biot had also noted that some species of quartz rotated polarized light in opposite directions, an observation that he was unable to explain after he had examined the crystal shapes. A typical crystal of quartz might be considered to have a shape approximating a hexagonal prism. Early in the century, the Abbé Haüy had observed the location of small facets at the top edges of quartz crystals that reduced the crystal symmetry. He noted that the placement of these facets (originally called 'plagihedral', later 'hemihedral') was such that the mirror image of the crystal would not be superimposable on itself — that is to say, there were 'left-handed' and 'right-handed' crystals. It was not until 1820 that John F. W. Herschel correlated the sign of rotation of plane-polarized light observed by Biot with the 'handedness' of the crystal. Herschel also observed that the chemical dissolution of an optically active piece of quartz destroyed the activity. Herschel concluded that greater attention should be paid to the presence or absence of hemihedral facets, a suggestion taken up by Pasteur some 25 years later in his studies of tartrate crystals.

Tartar is a crude form of potassium acid tartrate ($KHC_4H_4O_6$) that had been isolated as a deposit formed during wine fermentation. The free acid, tartaric acid ($C_4H_6O_4$), was first isolated by Carl Wilhelm Scheele (1742–1786) in 1769 and subsequently was produced commercially from tartar. In 1821 Charles Kestner, a manufacturer of chemicals in the Vosges in Alsace, obtained several hundred kilograms of a solid from a wine fermentation. Since the solid had different solubility properties from the tartar normally isolated, Kestner thought it might be oxalic acid. The supply obtained by Kestner was the only available to chemists for analysis for another 30 years. In 1828, Joseph Louis Gay-Lussac (1778–1850) determined that the new acid had the same chemical composition as tartaric acid and he accordingly named it 'acide racémique' (from the Latin *racemus*, grape). It was known as racemic acid in English. Berzelius used the name 'paratartaric acid' in 1832 and cited the existence of the two acids as an example of isomerism. In the discussions that follow, the names 'tartaric acid' and 'racemic acid' will be used in their original meaning. Upon the conclusion of the historical discussions that led to the understanding of the relationship between the two, the modern names will then be used.

In 1838, Biot reported that racemic acid was optically inactive, in contrast to tartaric acid which he had observed was optically active. Biot had attempted to ascertain the cause of the difference in properties but without success. Berzelius thought that the isomerism in the two acids might manifest itself in differences in the crystal structures of the salts. Since his studies were not very fruitful, he encouraged Eilhardt Mitscherlich (1794–1863) to pursue the study further since Mitscherlich had made considerable contributions to crystallography. By 1831, Mitscherlich had found that only the sodium-ammonium double salt of each had the same crystal form. Mitscherlich, however, did not publish the results of his investigations. He did not take any interest in the problem for another ten years.

In 1841, Frédéric Hervé de la Provostaye published a series of papers on crystallography. One of these was concerned with the crystal forms of tartaric acid and racemic acid. He found that in general the crystals of the various tartrate and racemate salts were not isomorphic. (The law of isomorphism had been formulated in 1819 by Mitscherlich. The law stated that the crystalline form of two substances should be the same if the number of combined atoms was the same. This would be expected to be

Plate 7. Louis Pasteur (1822–1895) in 1857 (Institut Pasteur).

the case even when the constituent elements were not identical. For example, crystals of the potassium and sodium salts should be isomorphic.) He had hoped to learn whether or not isomeric compounds should produce isomorphic crystals. In 1848, Louis Pasteur (1822–1895) (Plate 7), who was then an assistant to Biot, undertook a reinvestigation of de la Provostaye's work in order to give him some experience in the study of crystallography. He confirmed de la Provostaye's observations but with one important exception. He observed that the crystals of tartaric acid and its salts exhibited hemihedral facets. Racemic acid crystals, on the other hand, did not show this hemihedrism. Pasteur was therefore surprised to read a communication, published in 1844, in which Mitscherlich reported on the crystallographic identity of the sodium-ammonium salt of tartaric acid and racemic acid:

> The double paratartrate and the double tartrate of soda and ammonia have the same chemical composition $[(Na)(NH_4) C_4H_4O_6]$, the same crystalline form with the same angles, the same specific weight, the same double refraction, and consequently the same inclination in the optical axes. When dissolved in water their refraction is the same. But the dissolved tartrate deviates the plane of polarization, while the paratartrate is indifferent, as has been found by M. Biot for the whole series of those two kinds of salts. Yet, here the nature

and number of the atoms, their arrangement and distances, are the same in the two substances compared.

Pasteur had some difficulty in believing this conclusion since he assumed that the inactivity of the racemate salt should manifest itself in a crystal displaying no hemihedral facets. Pasteur prepared the sodium-ammonium salt and noted that not only did the tartrate contain the hemihedral facets, but to his surprise, the racemate as well. There was, however, a subtle difference between the two that had escaped Mitscherlich:

> Only the hemihedral facets in the tartrates all lay in the same direction; in the racemates some lay toward the right and some toward the left. In spite of the unexpected character of this result, I continued to follow up my idea. I carefully separated the crystals which were hemihedral to the right from those hemihedral to the left and examined their solutions separately in the polarizing apparatus. I then saw with no less surprise than pleasure that the crystals hemihedral to the right deviated the plane of polarization to the right, and that those hemihedral to the left deviated it to the left, and when I took an equal weight of each of the two kinds of crystals, the mixed solution was indifferent towards the light in consequence of the neutralization of the two equal and opposite individual deviations.

Thus, Pasteur had discovered that the cause of the optical inactivity of the racemic acid was due to it being a mixture of 'right-handed' and 'left-handed' tartaric acid. He had discovered the phenomenon of enantiomerism (or 'enantiomorphism' as it was called earlier) in the racemate crystals. Pasteur's success is even more remarkable when you consider that above approximately 26 °C (79 °F) both the tartrate and racemate crystallize as a monohydrate having no hemihedral faces or asymmetry.

When Biot was told of the result, he insisted that Pasteur repeat the experiment in his presence. After Biot had prepared the solution from the two kinds of crystals and had seen they were optically active, Pasteur related: 'The illustrious old man took me by the arm and said: "My dear child, I have loved science so much throughout my life that this makes my heart throb." '

Since the hemihedrism of the crystals is not always easily perceived, Pasteur prepared numerous crystal models to illustrate his lectures and also for Biot, whose eyesight was failing. Figure 17 shows two of Pasteur's models of the enantiomeric tartrate crystals.

The results of the studies undertaken by Pasteur in the next decade were no less remarkable. By the 1850s, he had discovered optical activity associated with a number of organic substances such as asparagine,

Figure 17. Pasteur's models of enantiomeric tartrate crystals (Institut
Pasteur).

aspartic acid and malic acid. He also observed that the isomeric compounds, fumaric acid and maleic acid, were optically inactive. A rather
interesting study was made by Pasteur of the crystals of strontium
formate which were observed to be hemihedral and optically active. A
solution of strontium formate, however, was optically inactive. He concluded that the hemihedrism of the crystal was

> . . . not due to the arrangement of the atoms in the chemical
> molecule, but to the arrangement of the physical molecules in
> the whole crystal, such that when the crystalline structure dis-
> appears in the act of dissolution there is no more dissymmetry.

This substance provides a second example (the other is quartz) of a
substance whose optical activity is observed only in the crystalline form
and on which can be observed the hemihedral facets. However, Pasteur
also found that the presence of hemihedrism in the crystalline form is not
always predictive of optical activity of solutions. For example, amyl
alcohol (2-methyl-1-butanol) obtained from certain natural sources is
optically active in solutions or as a liquid. Crystals of the barium
sulphamylate of the alcohol do not display any hemihedral facets. Thus

we can see that the absence or presence of hemihedrism in crystals cannot be used to predict the optical activity of the solutions. In this regard, Pasteur's investigation of malic acid, $HOOCH_2 CH(OH)COOH$, might have catalyzed the final stereochemical revolution that Wislicenus' study of lactic acid, $CH_3 CH(OH)COOH$, did a decade later. For a number of reasons, it did not.

In 1850, M. Dessaignes reported that he had prepared aspartic acid, $HOOCCH_2 CH(NH_2)COOH$, by the thermal decomposition of either ammonium malate or ammonium fumarate. Aspartic acid obtained from natural sources is optically active. Pasteur considered it inconceivable that a laboratory synthesis could produce an optically active substance, even a racemic mixture, from an optically inactive substance. The synthesis of aspartic acid from the optically active malic acid was acceptable to Pasteur, but it was difficult to believe that it might be prepared from fumaric acid. Pasteur obtained samples of the synthetic aspartic acid and determined that they were both optically inactive. Pasteur then converted some of the 'synthetic' aspartic acid to malic acid. On discovering that the malic acid was optically inactive, he concluded:

> Is it not evident that we have here to deal with a malic acid identical with the natural one [optically active], except for the simple suppression of its molecular dissymmetry? It is natural malic acid *untwisted*, if I may so express myself. The natural acid is a spiral stair as regards the arrangement of its atoms, this acid is the same stair made of the steps but straight in place instead of being spiral.
>
> It might be asked whether the new malic acid was not the racemic form of this group, that is, the compound of the right and left malic acids. This has very slight probability, for in this case not only would one active body be made from an inactive substance, but two active bodies would be produced, a dextro- and levorotatory substance.

Pasteur was thus convinced that it was impossible to synthesize an optically active substance — even if it was present in a racemic mixture. He therefore never attempted a chemical resolution (a procedure he developed about the same time) of the inactive product into its enantiomers. The fact that the crystals of malic acid do not exhibit hemihedral facets undoubtedly contributed to his thought that the malic acid was inactive due to its 'untwisted' symmetrical molecular form. We have seen, however, that the presence of crystal hemihedrism is not always essential to the prediction of molecular chirality.

However, Pasteur's views on the causes of the optical inactivity of malic acid led him to predict the existence of a second inactive form of tartaric acid. In 1853, Pasteur heated a solution containing tartaric acid and the alkaloid, cinchonicine. From the mixture he obtained not only racemic acid but also an optically inactive isomeric compound. From the discussion above, it is not difficult to understand Pasteur's surprise at finding racemic acid as a product – since its synthesis would involve a laboratory conversion of (+)-tartaric acid to (−)-tartaric acid. The optical inactivity of the new isomer was thought by Pasteur to arise from the 'untwisting' of ordinary tartaric acid. For some time this substance was known as 'inactive tartaric acid' in contrast to 'racemic acid'.

By the 1860s, Pasteur had classified the tartaric acid isomerism as follows:

1. *Tartaric acid*. The optically active form obtained from wine fermentation: *dextro*-tartaric acid (now designated (+)-tartaric acid).

2. *Racemic acid* (or paratartaric acid). The compound is composed of a 50:50 mixture of the enantiomeric tartaric acid: *dextro*- and *levo*-tartaric acid (now designated (±)-tartaric acid).

3. *Inactive tartaric acid*. This was Pasteur's 'untwisted' tartaric acid, now designated as *meso*-tartaric acid.

In the 1850s, Pasteur devised a more efficient method (chemical resolution) of separating the enantiomers in a racemic mixture. The method was developed as a consequence of his observation of the solubility differences of the salts of tartaric acid prepared from optically active alkaloids:

> When one prepares the racemate of cinchonicine, for example, it always occurs for a certain concentration of liquor, that the first crystallization is in major part formed of the *levo*-tartrate of cinchonicine; the *dextro*-tartrate remains in the mother liquor. Parallel results present themselves with quinicine, only, in this case, it is the *dextro*-tartrate which is first removed.

In short, Pasteur had discovered that he could take advantage of the solubility differences of diastereomeric salts as a method of separating enantiomers. In 1858, Pasteur also observed that a yeast mould growing in a solution containing the (±)-tartrate destroyed the (+)-tartrate leaving the (−)-tartrate.

The availability of these two methods (the 'chemical' and 'biochemical' methods) for obtaining an optically active substance was soon exploited by organic chemists. The reports of the isolation of increasing numbers of optically active substances in the 1860's coincides with the growing use of graphic formulas by organic chemists to illustrate the chemical structures of organic compounds. However, it was almost another twenty years before a relationship was perceived between the phenomenon of optical activity and molecular structure. Many chemists persisted in the belief that the observation of optical activity had little chemical significance. This belief had been stated sometime earlier (1841) by Charles Gerhardt (1816–1856):

> We chemists require chemical differences to distinguish between two bodies, and it therefore seems to me that those who attach such great importance to rotatory power are deluding themselves strongly if they look to it for the future of chemistry.

Although Gerhardt wrote this before Pasteur's work was published, the attitude persisted in the minds of most chemists for many decades. Even in the 1860s, some chemists described compounds that were identical in all respects except in this physical property as 'physical isomers'.

Pasteur himself was close to understanding the relationship between optical activity and molecular structure, but the final solution escaped him because he had turned to non-chemical studies in the 1850s and was therefore not in a position to appreciate the insights provided by the introduction of a theory of valency. This can be seen in a lecture given by Pasteur in 1860 in which he speculated on the reason why the dissolution of substances, that are optically active in the solid state, sometimes results in the loss of optical activity, other times in the retention of activity:

> Imagine a spiral stair whose steps are cubes, or any other objects with superposable images. Destroy the stair and the dissymmetry will have vanished. The dissymmetry of the stair was simply the result of the mode of arrangement of the component steps. Imagine, on the other hand, the same spiral stair to be constructed with irregular tetrahedra for steps. Destroy the stair and the dissymmetry will still exist, since it is a question of a collection of tetrahedra!

Pasteur then went on to consider the possible geometries of the molecular species that might produce the dissymmetry (asymmetry) that correlated with the observation of optical activity:

> Are the atoms of the right acid grouped on the spirals of a dextrogyrate helix, or placed at the summits of an irregular tetrahedron, or disposed according to some particular dissymmetric grouping or other? We cannot answer these questions.

But it cannot be doubted that there exists an arrangement of the atoms in a dissymmetric order, having a non-superposable image, and it is no less certain that the atoms of the *levo*-acid realize precisely the inverse dissymmetric grouping to this.

Pasteur's inattention to the developments in structural organic chemistry in the 1850s prevented him from considering the possible atomic arrangements of substances like tartaric acid that might produce the required 'dissymmetric groupings'. However, even the organic chemists who were aware of these developments considered Pasteur's proposals speculative inasmuch as there were difficulties enough in unraveling more conventional problems in organic chemistry. Only gradually did it become apparent that the phenomenon of 'optical isomerism' required a different theoretical interpretation than had been proposed to explain 'chemical' or 'structural' isomerism. The new interpretation arose from the experimental studies of the properties of a single compound, lactic acid.

Lactic acid had been discovered by Carl Wilhelm Scheele (1742–1786) in 1770 as a product isolated from sour milk. Although elemental analysis of the compound had been reported early in the 19th century, little was known of the chemical nature of the compound until the 1850s as the result of a controversy between Charles Wurtz and Hermann Kolbe. The basic difficulty revolved around the issue as to whether lactic acid should be considered 'monobasic' or 'dibasic'. To further complicate the issue was the report of the isolation of an acid from muscle tissue that had properties similar to lactic acid. This acid, referred to in the literature as 'sarcolactic acid' or 'paralactic acid', had the same chemical formula as lactic acid, but differed in its chemical and physical properties. By the early 1860s, those chemists concerned with the problem had concluded that the two acids were simply isomeric compounds: lactic acid corresponding to 2-hydroxypropanoic acid; sarcolactic acid, 3-hydroxypropanoic acid. One of the chemists who had become involved in the controversy was Johannes Adolf Wislicenus (1835–1902) (Plate 8). By 1863, he was persuaded that the two acids were not positional isomers. He concluded that the difference between the two could not be understood by the formulas then in use since they were limited to pictures of the substance in two dimensions. By the late 1860s, as the result of further experimental work, it was clear that the two acids were not positional isomers. But by this time, however, Wislicenus had discovered that whereas the substance known as lactic acid (isolated from sour milk) was optically inactive, sarcolactic acid was optically active. This observation was inexplicable to Wislicenus and suggested the existence of a new kind of isomerism:

Plate 8. Johannes Adolf Wislicenus (1835–1902).

Thus is given the first certainly proved case in which the number of isomers exceeds the number of possible structures. Facts like these compel us to explain different isomeric molecules with the same structural formula by different positions of their atoms in space and to seek for definite representations of these.

It is interesting to note that Kekulé merely summarized Wislicenus' remarks in a published account of the meeting at which Wislicenus spoke in 1869. Kekulé did not consider how his models might be used to solve the dilemma posed by Wislicenus:

The speaker drew our attention to the fact that the existence of three hydroxypropionic acids demonstrated the limitations of the structural formulas in general use, and also of the views that are usually expressed by these formulas. Such subtler cases of isomerism might perhaps be explained by the spatial representation of the combination of atoms, that is, by models.

Wislicenus continued to work on the problem for another four years but came no closer to understanding the cause of the differences. In 1873, he again concluded:

If molecules can be structurally identical and yet possess dissimilar properties, this can be explained only on the ground that the difference is due to a different arrangement of the atoms in space.

The new kind of isomerism was described by Wislicenus as 'geometrical isomerism' — a term that took on a different meaning a few years later. That Wislicenus did not attempt to arrange the 'atoms in space' with some sort of models might seem unusual to us now. However, the 'type' formulas used earlier by Wislicenus do not lend themselves to illustrations with models as well as the Crum Brown–Kekulé formulas (Figs. 10–12, 14):

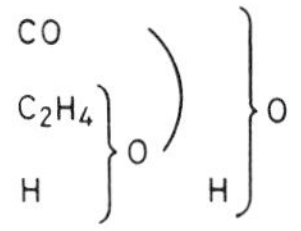

```
CO    ⎞ ⎞              CH₃
      ⎟ ⎟ ⎟            |
C₂H₄ ⎞ ⎟ ⎟ O          CH·OH
    ⎟ } O ⎟            |
H   ⎟   H ⎟            CO·OH
```

Type formula (1863) 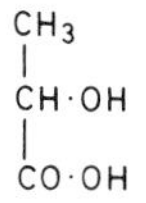   Graphic formula (1873)

By 1873, he was using graphic formulas which could have lent themselves to representations using the models developed by Kekulé. There is no evidence that he ever used these models, however. Perhaps Wislicenus did not wish to risk taking his time on such speculations when further experimental work was needed to establish for certainty that he was in fact observing a new kind of isomerism. As we have seen in the case of the paper published by Paternò, the literature contained a number of examples of reports of excessive isomers that later proved to be nonexistent.

It is also of interest to point out that the cause of the optical inactivity of milk-lactic acid was never dealt with. No chemical resolution was attempted (this was also the case with Pasteur's inactive malic acid). Thus, that the enantiomer of sarcolactic acid might be present in the inactive milk-lactic acid was never revealed. One can only conclude that Pasteur's observations in 1860, and earlier, had not had the profound effect on the 'geometrical' speculations of organic chemists that we might suppose from our present perspective.

5

The origins of stereochemistry: the contributions of J. H. van't Hoff and J. A. Le Bel

The solution to the questions posed by Pasteur and Wislicenus was provided in two papers published independently in 1874 by the Dutch chemist, Jacobus Henricus van't Hoff (1852–1911) and the French chemist, Joseph Achille Le Bel (1847–1930). Although both Van't Hoff and Le Bel had worked together in Wurtz's laboratory in Paris earlier in 1874, they apparently had not discussed the ideas that appeared in the papers. Le Bel's paper, published in the *Bulletin of the Chemical Society of Paris* in November 1874, was entitled 'On the Relations Which Exist Between the Atomic Formulas of Organic Compounds and the Rotatory Power of their Solutions'. Van't Hoff's ideas appeared in a 12-page booklet in Dutch in September 1874 which was entitled 'A Suggestion Looking to the Extension into Space of the Structural Formulas at Present Used in Chemistry – and a Note upon the Relation between the Optical Activity and the Chemical Constitution of Organic Compounds'. Although a French translation of this booklet appeared shortly afterwards, Van't Hoff's ideas were more widely disseminated after the publication of an expanded (44 pages) version in 1875, the title of which was *La Chimie dans l'Espace*. A German translation was published in 1877.

Although Van't Hoff (Plate 9) had worked in Kekulé's laboratory at Bonn for about a year in 1872–73, he does not mention Kekulé's tetrahedral carbon models. In 1894, Van't Hoff stated:

> On the whole, Le Bel's paper and mine are in accord; still, the conceptions are not quite the same. Historically the difference lies in this, that Le Bel's starting point was the researches of Pasteur, mine those of Kekulé. My conception is a continuation of Kekulé's law of the quadrivalency of carbon, with the added hypothesis that the four valences are directed toward the corners of a tetrahedron, at the center of which is the carbon atom.

Plate 9. Jacobus Henricus van't Hoff (1852–1911) (Edgar Fahs Smith Collection).

Although Van't Hoff acknowledged his debt to Kekulé, it was only to Kekulé's theory of the tetravalency of carbon and not to his geometrical extension of the theory as illustrated in the tetrahedral models. Thirty years later, Van't Hoff recounted how his reading of Wislicenus' 1874 paper on lactic acid provided him with the inspiration for the tetrahedral carbon atom:

> Students, let me give you a recipe for making discoveries. In connection with what has just been said about libraries, I might remark that they have always had a mind-deadening effect on me. When Wislicenus' paper on lactic acid appeared and I was studying it in the Utrecht library, I therefore broke off my study half-way through, to go for a walk; and it was during this walk, under the influence of the fresh air, that the idea of the asymmetric carbon atom first struck me.

Van't Hoff began his paper:

> It appears more and more that the present constitutional formulae are incapable of explaining certain cases of isomerism; the reason for this is perhaps the fact that we need a more definite statement about the actual positions of the atoms.

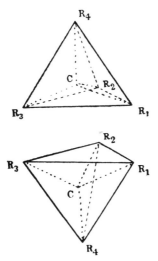

Figure 18. Van't Hoff's illustration of the enantiomers of $CR_1R_2R_3R_4$.

He first pointed out that if the four affinities of carbon lay in a plane at right angles to each other, an excess number of isomeric derivatives of methane is predicted. For example, for a disubstituted methane (CH_2R_2, $CH_2R'R''$, CHR_2R'') the number should be two; for trisubstituted derivatives ($CHR'R''R'''$ or $CR'R''R'''R''''$), three. (You may recall that Victor Meyer had made the same observation a few years earlier.) The tetrahedral carbon concept reduces the number of isomers to one, except in the last case where the four substituents are all different:

> The theory is brought into accord with the facts if we consider the affinities of the carbon atom directed toward the corners of a tetrahedron of which the carbon atom itself occupies the center. When the four affinities of the carbon atom are satisfied by four univalent groups differing among themselves, two and not more than two different tetrahedrons are obtained, one of which is the reflected image of the other, they cannot be super-imposed; that is, we have here to deal with two structural formulas isomeric in space.

The latter mirror-image isomers (*enantiomers*) were illustrated in the article (Fig. 18).

Carbon atoms which contained four different substituents were referred to as 'asymmetric carbon atoms'. The presence of such asymmetric carbon atoms in a molecule could be used to predict whether a substance might be optically active:

. . . all of the compounds of carbon which in solution rotate the plane of polarized light possess an asymmetric carbon atom.

The observation of optical inactivity indicated either the absence of such asymmetric carbons or that the substance existed as a 50:50 mixture of the enantiomers (a racemic mixture). In the 1874 booklet, Van't Hoff does not discuss the optical inactivity of compounds (such as the 'inactive' tartaric acid) containing more than one asymmetric carbon atom. Marcellin Berthelot (1827–1907) in fact criticized Van't Hoff (and Le Bel) for not accounting for the optical inactivity of the 'indivisible inactive type' of compounds, such as the inactive forms of tartaric acid and malic acid reported by Pasteur. Van't Hoff responded to this criticism in a booklet published in 1875. The maximum number of stereoisomers possible was shown by Van't Hoff to equal 2^n, where n = the number of asymmetric carbon atoms. In the case of compounds containing two asymmetric carbon atoms, each of which had the same substituents, two of the mirror-image isomers would be identical, and therefore that particular compound would be optically inactive. This is illustrated by the general formula, $C(R_1 R_2 R_3)C(R_1 R_2 R_3)$, which in the case of tartaric

Figure 19. Cardboard tetrahedral models prepared by J. H. van't Hoff in 1874–75 (History of Science Museum, Leiden).

acid, the three substituents (R_1, R_2 and R_3) correspond to H, OH and COOH.

Van't Hoff realized that the relationships between the various stereo-isomers were not always easily visualized — even in the perspective drawings included in the publication. He therefore prepared a number of cardboard models (Fig. 19) which he sent along with a copy of the 1875 booklet to a number of well-known chemists: A. von Baeyer (Strasbourg), A. Butlerov (St Petersburg), L. Henry (Louvain), A. Hofmann (Berlin), A. Kekulé (Bonn), E. Frankland (London), J. Wislicenus (Würzburg), A. Wurtz and M. Berthelot (Paris).

Six of the models shown here consist of single tetrahedra (about 2 cm on edge) that illustrate the enantiomers of malic acid and tartaric acid (with respect to only one of the two asymmetric carbon atoms). One model illustrates the symmetry found in malonic acid, $HOOCCH_2 CH_2 COOH$, with respect to one of the methylene carbons. The remaining three models consist of two tetrahedra joined at their faces and illustrate (+)- and (−)-tartaric acid and *meso*-tartaric acid. In these models the carbon valencies pass through the *faces* of the tetrahedra rather than the apexes as is more common (see, for example, Fig. 18). Normally two tetrahedra that share a face represent the carbon—carbon triple bond (see the discussion below). The collection of the Van't Hoff models preserved in the Deutsches Museum in Munich contains a variety of models that were used by Van't Hoff to illustrate his solution to a number of stereochemical problems.

The models in Fig. 19 constructed by joining two tetrahedra faces represent particular 'conformations' of the tartaric acid stereoisomers. The term 'conformation' was not used by Van't Hoff — instead he talks about the various possible 'phases' of the molecule, which should not be considered in the same sense as 'isomers'. It was only in a later edition of his book (1891) that he considered whether there might be preferred 'phases':

> To avoid the prediction of an isomerism [for a substance such as $C(R)_3 C(R_1)_3$] which at first sight appears endless, it is not necessary to introduce any additional hypothesis; the difficulty disappears at once if we take into account the mutual action which must take place between the groups R_3 and the groups R_1, united to each of the two atoms of linked carbon. In fact, if this action depends, as in the case with every known force besides, on the distance and nature of the groups in question, there will be among the possible positions only one which corresponds to the state of stability.

Later (1898) Van't Hoff suggested that there must be a 'free rotation' about the carbon–carbon single bonds since not to assume that would allow for an excess of isomers. Nevertheless, he thought that the interactions of the substituents should lead to a 'favored configuration'.

The second half of Van't Hoff's 1874 booklet was entitled 'The Influence of the New Hypothesis upon Compounds Containing Doubly Linked Carbon Atoms'. The theoretical implications of the proposals contained in this section had an importance perhaps equal to that of the tetrahedral carbon atom. The proposal provided a solution to the observation of several puzzling cases of isomerism – that found, for example, between maleic acid and fumaric acid. Although it was not clear to many chemists, such as Kekulé, that these compounds were not structural isomers, Van't Hoff felt the evidence indicated that these compounds were 'geometrical isomers' (a term used by Wislicenus somewhat earlier). Van't Hoff's discussion below refers to the two illustrations shown in Fig. 20:

> Double linking is represented by two tetrahedrons with one edge in common in which A and B represent the union of the two carbon atoms, and $R_1 R_2 R_3 R_4$ represent the univalent groups which saturate the remaining free affinities of the carbon atoms. If $R_1 R_2 R_3 R_4$ all represent the same group, then but one form is conceivable, and the same is true if R_1 and R_2 or R_3 and R_4 are identical, but if R_1 differs from R_2 and at the same time R_3 differs from R_4, which does not preclude R_1 and R_3, R_2 and R_3 from being equal, then two figures become possible, which differ from one another in regard to the positions of R_1 and R_2 with respect to R_3 and R_4. The dissimilarity of these figures, which are limited to two, indicates a case of

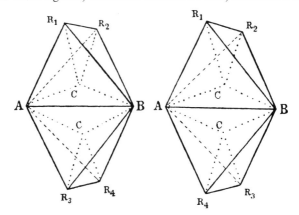

Figure 20. Van't Hoff's illustration of geometrical isomers (1874).

isomerism not shown by the ordinary formulas. [To illustrate the maleic acid/fumaric acid isomerism: R_1, R_3 = H and R_2, R_4 = COOH.]

Since the double bond involved the sharing of the edges of two tetrahedra, it was a natural extension for Van't Hoff to consider that the triple bond involved the sharing of the faces of two tetrahedra (Fig. 21). There was, however, no experimental evidence available that could be used to support this suggestion.

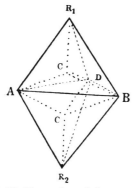

Figure 21. Van't Hoff's illustration of the structure of an alkyne, $R_1C{=}CR_2$.

The style and scope of the paper published by Le Bel in 1874 differs significantly from Van't Hoff's. Although the names of both of these men are associated with the origins of stereochemistry, an historical analysis would suggest that Van't Hoff's proposals had the greater impact on the subsequent development of stereochemistry. Le Bel (Plate 10) was more concerned with understanding the relationship between molecular symmetry and the observation of optical activity:

> The labors of Pasteur and others have completely established the correlation which exists between molecular asymmetry and rotatory power. If the asymmetry exists only in the crystalline form the crystal alone will be active; if, on the contrary, it belongs to the chemical molecule the solution will show rotatory power, and often the crystal also if the structure of the crystal allows us to perceive it, as in the case of the sulphate of strychnine and the alum of amylamine.

In contrast to Van't Hoff, the concept of the tetrahedral carbon atom is considered only once by Le Bel, and then only in the context of a general discussion of molecular geometry:

Plate 10. Joseph Achille Le Bel (1847–1930) (Edgar Fahs Smith Collection).

Let us consider a molecule of a chemical compound having the formula MA_4; M being a single or complex radical combined with four univalent atoms A, capable of being replaced by substitution. Let us replace three of them by simple or complex univalent radicals differing from one another and from M; the body obtained will be asymmetric. Indeed, the group of radicals R, R', R'', A when considered as material points differing among themselves form a structure which is enantiomorphous with its reflected image, and the residue M cannot re-establish the symmetry.

Again, if it happens not only that a single substitution furnishes but one derivative, but also that two and even three substitutions give only one and the same chemical isomer [that is, other than optical isomers], we are obliged to admit that the four atoms A occupy the angles of a regular tetrahedron, whose planes of symmetry are identical with those of the whole molecule MA_4; in this case also no bisubstitution product can have rotatory power.

Unlike Van't Hoff, Le Bel does not use perspective drawings or models to illustrate his ideas.

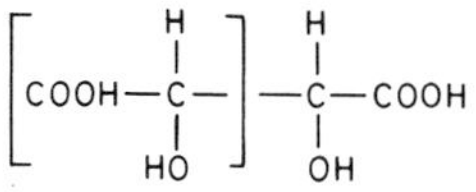

$$\left[COOH-\underset{\underset{HO}{|}}{\overset{\overset{H}{|}}{C}}- \right] -\underset{\underset{OH}{|}}{\overset{\overset{H}{|}}{C}}-COOH$$

The cause of the isomerism in tartaric acid was discussed in terms of a formula (shown above) which can be derived by replacing three of the hydrogens in methane by HO, COOH and COOH–CH(OH):

> Moreover, an examination of the formula shows that the last of the substituted groups is identical with the grouping of the entire remainder of the compound; we have to deal therefore, with the second class of exceptions to the first general principle . . . if the two groups combined with one another are identical and superposable their effect upon polarized light will be added — this is what takes place with the active acid; if, on the contrary, the combined groups have an inverse symmetry they will exactly neutralize one another and we will have the inactive tartaric acid.

That the tetrahedral carbon atom did not play a central role in Le Bel's concept of the geometry of molecules can be seen from his treatment of unsaturated compounds. Since Van't Hoff's structure of substituted alkenes (Fig. 20) is symmetrical, optical activity for this class of compounds is impossible. Le Bel, on the other hand, felt that the four substituents on the carbons of the carbon–carbon double bond were not coplanar:

> . . . To explain the isomerism of the ethylene derivatives, we must suppose the hydrogen atoms to be at the angles of a hemihedral quadratic pyramid superposable upon its image . . . and we should obtain by two substitutions two isomers, one of which would be symmetrical, and the other asymmetrical. These isomers will both be symmetrical if the two substituted radicals are the same, as happens in the case of maleic and fumaric acids. Hence it is sufficient for the optical study of two disubstitution derivatives, such as the amylene of active amyl alcohol, $CH = C\overset{CH_3}{\underset{C_2H_5}{}}$, and its isomer $CH_3 CH = CH-C_2 H_5$, to decide whether the four hydrogen atoms are in the same plane or not.

Several things need to be clarified in this passage. First of all, Le Bel apparently assumed that the formulas for maleic acid and fumaric acid

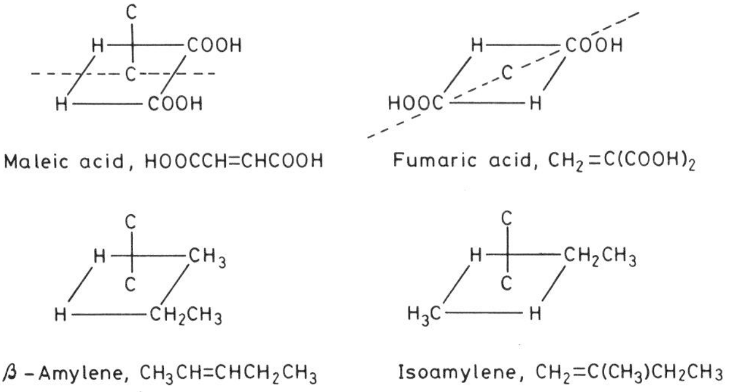

Figure 22. Probable structures of ethylene derivatives according to Le Bel.

were $HOOCCH = CHCOOCH$ and $CH_2 = C(COOH)_2$, a view not shared by Van't Hoff. Even when Le Bel later corrected this view, he still had not abandoned the geometrical structure he had proposed for substituted ethylenes. Although Le Bel does not include any drawings to illustrate what was meant by a 'hemihedral quadratic pyramid', the best guess is that he was talking about a square pyramid (Fig. 22).

If these in fact correspond to the structures envisioned by Le Bel, it can be seen that neither maleic acid nor fumaric acid can be expected to exist in optically active form, since both contain a plane of symmetry. (The dotted line indicates where the plane would bisect the base of the square pyramid.) Isoamylene can also be expected to be inactive. On the other hand, the mirror image of β-amylene is non-superimposable on itself and could therefore be obtained in an optically active form. Even though Le Bel's attempts to confirm his theory experimentally were without success, he was reluctant to abandon his ideas even 15 years later. Nor did Le Bel wish to have his name associated with the concept of the tetrahedral carbon atom, as can be seen from statements he made repeatedly in the early 1890s:

> I use the greatest efforts in all my explanations to abstain from basing my ideas on the preliminary hypothesis that compounds of carbon of the formula CR_4 have the shape of a regular tetrahedron.

Le Bel considered that the particular geometry would be determined by the 'zones of repulsions' surrounding each atom in the molecule. It was possible then for a methane derivative, CR_4, to have a tetragonal pyramidal structure (similar to the square pyramid, but without the second carbon atom in the base). Although Le Bel was unsuccessful in his search for the observation of optical activity in a number of unsaturated compounds, his studies of quaternary ammonium compounds initiated the development of inorganic stereochemistry (Chapter 10).

Van't Hoff's demonstration in 1874 that all of the known optically active compounds contained an 'asymmetric' carbon atom provided organic chemists with a convenient means by which they could decide whether optical isomerism was possible for a new compound. The concept did not take an immediate foothold, however, because of reports of the isolation of optically active compounds which did not contain any asymmetric carbon atom. Among some of the compounds initially reported were: styrene, 1-propanol, α-picoline, papaverine, chlorofumaric acid and chloromaleic acid. Within a decade or so, it was demonstrated that the optical activity associated with these compounds was due to the presence of optically active impurities. Any reservations chemists had about Van't Hoff's proposal were removed by the reports of increasing numbers of optically active substances, all of which were found to contain asymmetric carbon atoms. In the late 1880s a few chemists, such as A. C. Brown and P. A. Guyé, undertook experimental and theoretical studies designed to elucidate the relationship between the sign (and magnitude) of the optical rotation and the nature of the groups bound to the asymmetric carbon atom. It was not always clear, however, whether optical activity would be observed in molecules having a small number of carbon atoms – since some felt that the stability of the asymmetric group was controlled by the bulk of the attached groups. Le Bel, for example, thought that the tetrahedral arrangement would be preferred only when the four groups were of sufficient size to exert repulsive forces that would favor that geometry over a square pyramid. Another chemist, Arnold Eiloart, proposed in 1898 that optically active compounds should contain at least three carbon atoms, since smaller compounds could undergo 'intramolecular transformations favored by the mobility of the small radicals attached to the asymmetric carbon'. The experimental refutation of this idea, however, was not provided until 1914 by W. J. Pope and J. Read who reported the chemical resolution of a compound containing only one carbon atom: chloroiodomethanesulfonic acid, $ClIHCSO_3H$.

When Van't Hoff's and Le Bel's papers were published in 1874, few chemists considered that speculations about the arrangement of atoms in space would be of any practical value for organic chemical research. From most of the chemists who were sent reprints of the 1875 pamphlet and the cardboard models, Van't Hoff received only a polite response. Adolph von Baeyer, however, was enthusiastic and showed the book and the models to his students in his laboratory with the remark: 'Here we have again a really new good thought in our science, which will bear ripe fruit.'

Wislicenus, as can be imagined after working so many years on the lactic acid problem, was quite receptive to Van't Hoff's proposals and was instrumental in seeing to the publication of an expanded German

Plate 11. Hermann Kolbe (1818–1884).

edition of Van't Hoff's book. The translation, published in 1877, contained a preface in which Wislicenus urged chemists to consider the ideas of Van't Hoff seriously, in spite of what he recognized was the scepticism of most chemists of that time:

> It is no long time since strong protests were frequently made by champions of the most advanced views in theoretical chemistry, against the idea that chemistry would ever reach such a point as to bring forward the conception of the position of the atoms in space in order to explain the properties of a compound.

One of the more influential chemists of the period, and one who was highly suspicious of the theoretical speculations of many chemists, was Adolf Wilhelm Hermann Kolbe (1818–1884) (Plate 11). In his position as editor of the *Journal für Praktische Chemie*, he often took it upon himself to comment on the deficiencies of chemical research. Kolbe attributed the decay in chemical research in Germany to the introduction of an increasing amount of 'speculative philosophy' which indicated a 'lack of general and at the same time thorough chemical training' in chemical research:

> If anyone supposes that I exaggerate this evil, I recommend him to read, if he has the patience, the recent fanciful publications of Messrs Van't Hoff and Hermann on *Die Lagerung der Atome*

in Raume. This paper like many others, I should have been content to ignore had not a distinguished chemist taken it under his protection and warmly recommended it as a performance of merit.

A Dr J. H. van't Hoff, of the Veterinary College, Utrecht, appears to have no taste for exact chemical research. He finds it a less arduous task to mount his Pegasus (evidently borrowed from the Veterinary College) and to soar to his Chemical Parnassus, there to reveal in his *La Chimie dans l'Espace* how he finds the atoms situated in the world's space.

It is not possible, even cursorily, to criticize this paper, since its fanciful nonsense carefully avoids any basis of fact, and is quite unintelligible, to the calm investigator . . . It is one of the signs of the times that modern chemists hold themselves bound and consider themselves in a position to give an explanation for everything, and when their knowledge fails them to make use of supernatural explanations. Such a treatment of scientific subjects, not many degrees removed from a belief in witches and from spirit-rapping, even Wislicenus considers permissible.

As Van't Hoff commented ten years later:

Such was the debut of this theory. Only fourteen years have passed, Kolbe is now dead, and, as if by the irony of fate his place at the University of Leipzig has been taken by Wislicenus.

If anything, Kolbe's intemperate attack probably did more than anything Van't Hoff or Wislicenus could have done to publicize the proposals. Kolbe probably represents an extreme view of the scepticism that many chemists had for theoretical speculations. It should be remembered that the atomic theory itself was considered only a useful hypothesis a decade earlier, as can be seen, for example, in Kekulé's comments in 1867:

. . . The question whether atoms exist or not has but little significance in a chemical point of view, its discussion belongs rather to metaphysics. In chemistry we have only to decide whether the assumption of atoms is an hypothesis adapted to the explanation of chemical phenomena.

It should come as no surprise then to learn that Kekulé was only mildly receptive to Van't Hoff's ideas and referred to the tetrahedral carbon atom as a 'usable hypothesis'.

There were, however, more legitimate sources of criticism of Van't Hoff's ideas, mainly from the physical chemists who had difficulty in accepting Van't Hoff's proposal of spatially directed valence forces. As A. Claus noted in 1881:

> . . . to suppose that the chemical attraction inherent in its atom is divided into parts each of which acts independently of each other is as unnatural as it is unfounded.

Claus argued that the total, unified affinity of carbon only separated into parts when it combines with other atoms. Most organic chemists, however, were not affected by this kind of criticism since they found the idea of the tetrahedral carbon atom useful to explain the existence of stereo-isomers. That it violated the physics of the period did not concern them to any great degree. The 1880s was a period in which a number of physical and organic chemists engaged in polemics directed toward decid-ing on the 'shape' of the carbon atom. To summarize the major points of the discussion: Should the carbon atom be regarded as a sphere with the four valencies oriented towards the corners of a tetrahedron, or did the atom itself have a tetrahedral shape with the valency forces concentrated on either the corners or faces? As Demut and Meyer stated in 1881:

> We cannot assume that valences come across in the empty space free of atoms, it is only possible on paper or in a model where these are lines or wires but not forces.

While Van't Hoff himself rather generally supposed that the number and direction of the valences should be related to the shape of the atoms, Wislicenus thought it possible that the carbon atom might be tetrahedral in shape with the maximum attraction concentrated at the apexes of the tetrahedron. Other chemists (Wunderlich, Knoevenagel, Knorr, Vaubel and Auwers) proposed various models in which the atom was considered to have several points of attraction whose number and location were dependent upon the shape of the atom. H. Sachse (Chapter 13) even used the tetrahedral shape of the carbon atoms as the basis for his calculations of what he thought would be the relative carbon–carbon single, double and triple bond lengths.

In 1891 Alfred Werner discussed the stereochemistry of carbon com-pounds in terms of his new theory of valency (see Kauffman's book, *Inorganic Chemical Complexes*, in this series). Although the attractive power of an atom is distributed evenly over its surface, in the case of carbon, the number of atoms it can attract is limited to four and these will be distributed over the surface so as to produce the greatest possible neutralization of the reciprocal affinities between the carbon atom and the groups. If the four groups were identical, this would presumably produce a tetrahedral geometry. Not all of the affinity of the carbon atom is completely neutralized, however, and can therefore be used to form molecular complexes which could be either relatively stable or transitory – as would be illustrated in the Walden inversion (Chapter 7). The utilization of the residual affinity of a centrally binding atom had its greatest success in explaining the existence and stereochemistry of inorganic complexes (Chapter 7).

In unsaturated compounds, Werner thought that some of the 'excess' of free affinity not utilized in the formation of the carbon–carbon 'double' bond, was oriented in such a way as to restrict rotation about the bond. The tendency of the atoms in cyclic compounds to return to positions which produced the greatest possible neutralization of affinities provided a qualitatively appealing explanation of Baeyer's 'tension' theory (Chapter 13). Werner's theory contrasted with that of Le Bel (1890) who argued that the zone of repulsion about the atoms determined the molecular geometry.

Most of the theories would be consistent with the existence of irregular tetrahedral geometries in compounds in which the four groups joined to a carbon atom were not identical. This was first suggested by Van't Hoff himself in 1877 and his book includes diagrams illustrating the construction of irregular tetrahedra from cardboard. It was, of course, difficult to experimentally establish whether in fact the valences were 'equivalent' or not, but a few attempts were made to answer this question. In 1887, L. Henry could find no differences between the two 'isomeric' (α- and β-) acetonitriles produced in a series of substitution reactions:

$$CH_3 I + KCN \longrightarrow CH_3 - CN \longrightarrow CH_3 COOH \longrightarrow ClCH_2 COOH \longrightarrow NC-CH_3$$
$$(\alpha -) \qquad\qquad\qquad\qquad\qquad\qquad\qquad\qquad (\beta -)$$

He concluded, therefore, that the valency forces themselves were equivalent. That this work was not without ambiguity can be found in the observation of the abstractor of the paper:

The whole argument is, of course, based on the principle of substitution and the stability of complex molecular structure throughout the course of a chemical reaction.

This synthesis was done some ten years before Walden's observations (Chapter 7) called into question the assumptions that had previously been made regarding the stereochemistry of substitution reactions, which reopened the debate. For example, in 1914 Emil Fischer re-examined some

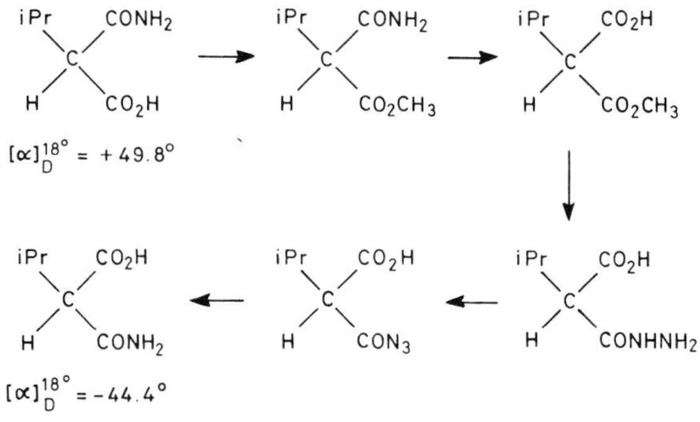

Figure 23. E. Fischer's experimental demonstration of the interconversion of enantiomers produced by the interchange of two groups attached to an asymmetric carbon.

of his assumptions regarding the geometry and stability of the asymmetric carbon atom, even though this had been assumed in his earlier work in establishing the configurations of the carbohydrates. Would the interchange of two groups (not involving the bonds to the asymmetric carbon) on an asymmetric carbon atom produce the enantiomer? The series of reactions (Fig. 23) studied by Fischer provided him with an affirmative answer to the question.

Ten years later, however, K. Weissenberg and H. Mark called into question the primacy of the tetrahedral carbon atom with their interpretation of the X-ray crystallographic study of pentaerythritol, $(HOCH_2)_4C$. They proposed that the central carbon atom was to be found at the apex of a square pyramid. Suffice it to say, this proposal generated considerable interest and controversy for several years until it was determined that the original data were found to be in error.

Although there continued to be other isolated reports (for example by A. N. Campbell in the 1930s relating to the non-identity of enantiomers) that questioned the concept of the tetrahedral carbon, the rapid rise of stereochemical investigations in the latter part of the 19th century attests to the general acceptance of the main assumptions of the theory. Where difficulties were found, they arose in areas in which the theory had been extended in order to explain the stereochemistry of compounds not containing asymmetric carbons or the stereochemistry of substitution, addition or elimination reactions.

It was not until the 1930s that a satisfactory theory was available that explained directed 'valences'. This was provided by Linus Pauling in 1931 based on a quantum-mechanical interpretation of the chemical bond. Pauling calculated the eigenfunctions for an sp^3-hybridized atom and concluded:

We have thus derived the result that an atom in which only s and p eigenfunctions contributes to bond formation and in which quantization in polar coordinates is broken can form one, two, three, or four equivalent bonds, which are directed toward the corners of a regular tetrahedron. This calculation provides the quantum mechanical justification of the chemist's tetrahedral carbon atom.

This represented a considerable theoretical advance over the ideas proposed by G. N. Lewis some ten years earlier (see Chapter 10). More recently, a number of chemists have revived and extended the Lewis–Langmuir theory with considerable success in predicting the geometry of molecules. The 'valence shell electron pair repulsion' theory that now enjoys considerable popularity was largely the result of the proposals of Ronald Nyholm and Ronald J. Gillespie in the late 1950s. The latter chemist has been responsible for the popularization of the ideas, although J. Linnett has also contributed to the utility of the approach.

6

The stereochemistry of addition and elimination reactions

In 1887 Wislicenus published a pamphlet entitled *The Space Arrangements of the Atoms in Organic Molecules and the Resulting Geometrical Isomerism in Unsaturated Compounds.* Wislicenus felt that the theories of Van't Hoff and Le Bel had made few advances except those which

> . . . consist in the proof of the fact that optically active organic compounds of known structure always contain an asymmetric carbon atom, and that the giving up of asymmetry is at once accompanied with a loss of optical activity, and that bodies which are optically inactive in spite of the presence of an asymmetric carbon atom, are mixtures or compounds of oppositely active modification . . . [however] the idea announced by Van't Hoff and Le Bel, concerning the isomerism of certain unsaturated compounds, such as fumaric acid and maleic acid or crotonic and isocrotonic acid, has, on the contrary, remained unfruitful up to the present time.

Wislicenus attempted to remedy this situation by not only explaining a number of puzzling cases of chemical isomerism in stereochemical terms but also showing how the stereochemistry of a reaction might be predicted. Most of the discussion centered around the reactions involving maleic acid and fumaric acid. The molecular geometry proposed by Van't Hoff (Fig. 20) did not allow chemists to predict the stereochemistry of addition reactions (such as with bromine, Br_2), which produced different products from the two acids. From these observations, Kekulé and Rudolf Fittig had earlier argued that the two acids, as well as the products, must be structural isomers. It was suggested that in maleic acid the two 'unsatisfied' valences were on the same carbon. The saturation of these two valences by bromine would therefore produce a different product than

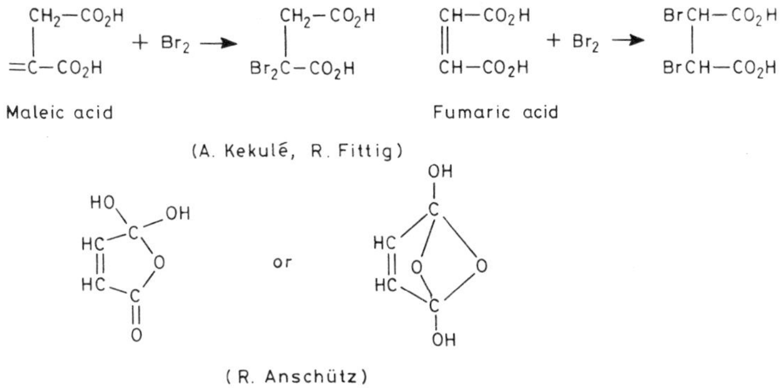

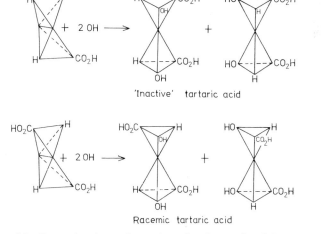

$$
\begin{array}{ccc}
\underset{\underset{\displaystyle =C-CO_2H}{|}}{CH_2-CO_2H} & +\ Br_2 \longrightarrow & \underset{\underset{\displaystyle Br_2C-CO_2H}{|}}{CH_2-CO_2H}
\end{array}
\qquad
\begin{array}{ccc}
\underset{\underset{\displaystyle CH-CO_2H}{\|}}{CH-CO_2H} & +\ Br_2 \longrightarrow & \underset{\underset{\displaystyle BrCH-CO_2H}{|}}{BrCH-CO_2H}
\end{array}
$$

Maleic acid   Fumaric acid

(A. Kekulé, R. Fittig)

or

(R. Anschütz)

Figure 24. Structures of maleic and fumaric acid proposed before 1874.

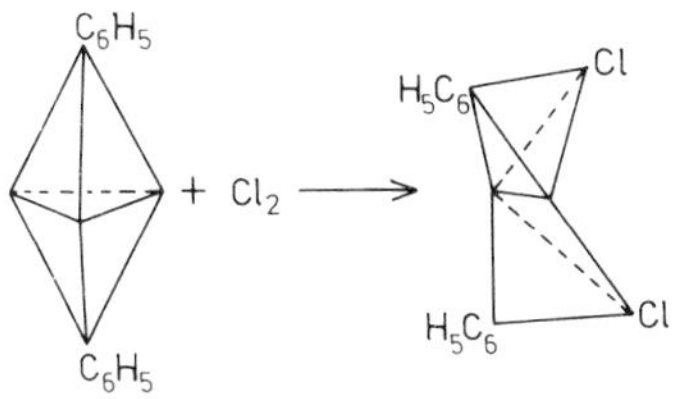

Figure 25. Postulated '*cis*' addition of Cl_2 to $C_6H_5C{\equiv}CC_6H_5$.

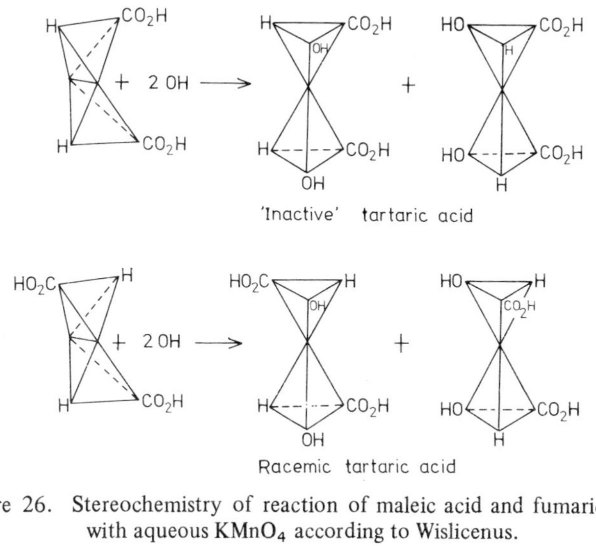

'Inactive' tartaric acid

Racemic tartaric acid

Figure 26. Stereochemistry of reaction of maleic acid and fumaric acid
with aqueous $KMnO_4$ according to Wislicenus.

from fumaric acid (Fig. 24, top). Richard Anschütz, on the other hand, suggested that maleic acid had a cyclic structure (Fig. 24, bottom).

Wislicenus assumed that in the *addition* reaction involving double or triple bonds, the reagent (such as Br_2) must add from only one side of the unsaturated linkage. For example, in the reaction of chlorine with diphenylethyne, $C_6H_5C \equiv CC_6H_5$, only the *cis*-dichloro derivative, $C_6H_5C(Cl) = C(Cl)C_6H_5$, should be formed (Fig. 25). Similarly, the re-action of maleic and fumaric acid with aqueous $KMnO_4$ should involve the *cis* addition of two hydroxyls to give the 'inactive' and 'racemic' tartaric acid, respectively (Fig. 26).

The assignment of the spatial positions of the carboxyl groups in *cis/trans* isomers such as maleic and fumaric acid was often based on the ease with which they underwent intramolecular cyclization reactions. The observation that maleic acid formed a cyclic anhydride on heating was used by Van't Hoff to support his view that the two carboxyl groups were closer together than in fumaric acid, which did not form a cyclic an-hydride. This experimental test, however, was not always reliable and sometimes led to incorrect structural assignment for other dicarboxylic acids.

The stereochemistry of *elimination* reactions was somewhat more complicated. Although Wislicenus assumed that the eliminated groups would also leave from the same side of the molecule, this assumption required that 'conformational' problems be considered. Even though Van't Hoff had assumed that there might be a 'favored' molecular con-formation (which was determined by the nature of the attractive/repulsive forces between non-bonding groups), the elimination might require the use of a 'less favorable' conformation that would place the eliminated groups on the same side of the molecule. Thus the formation of maleic acid from malic acid at high temperatures was thought to require the elimination of water from malic acid in a less favorable conformation (Fig. 27).

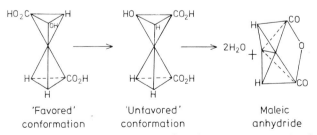

'Favored'   'Unfavored'   Maleic
conformation   conformation   anhydride

Figure 27. Presumed stereochemistry of an elimination reaction involving an 'unfavorable' conformation.

The HBr-catalyzed isomerization of maleic acid to fumaric acid was explained in a similar manner. When HBr adds to one side of the double bond, a 'less favorable' conformation of 2-bromosuccinic acid is formed. Rotation about the carbon–carbon bond produced a more stable conformation from which the elimination of HBr produces fumaric acid.

By the early 20th century most chemists were generally persuaded by Wislicenus' arguments to take it as axiomatic that the reactions involving the formation/reaction of double or triple bonds required the simultaneous elimination/addition from only one side of the molecule. The few experimental observations that had been made that seemed inconsistent with this stereochemistry were attributed to an incomplete understanding of the factors that determine the conformation in which the molecule reacted.

The experimental studies initiated in the 1890s by Arthur Michael, at Tufts College in Massachusetts, furnished the most serious challenge to the concepts of Van't Hoff and Wislicenus. Michael was able to cite numerous examples of reactions whose stereochemistry was not consistent with *cis* addition or elimination. For example, he observed that when hydrogen halides reacted with acetylene dicarboxylic acid, $HOOC-C\equiv C-COOH$, the product was the 2-halofumaric acid rather than the 2-halomaleic acid that would be expected for a *cis* addition of HX to the triple bond.

Michael was not content merely to challenge the experimental basis of some of the assumptions and extensions of Van't Hoff's theory. Instead he rejected the theory itself, thinking that its major deficiency lay in its inability to distinguish between what he designated as 'physical' and 'chemical' isomerism. The differences in the physical properties of isomeric molecules were described by Michael by the term 'alloisomerism'. Van't Hoff could only point out in response that 'Michael's hypothesis is unassailable because it offers not an explanation but an expression of the observed phenomena'. Michael's attack on Van't Hoff's theory perhaps prevented chemists from considering more seriously the questions raised by his experimental studies. The evidence that questioned the stereochemistry of addition and elimination reactions continued to accumulate, however. A particularly troublesome series of reactions that had been observed is summarized in Fig. 28.

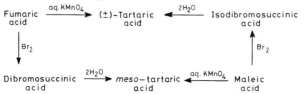

Figure 28.   Stereochemistry of fumaric acid–tartaric acid–maleic acid interconversions.

If, for example, the reaction of fumaric acid with bromine (step 1) involves a *cis* addition, and if the subsequent hydrolysis (step 2) also involves a *cis* replacement of the two bromines, it can be expected that the (±)-tartaric acid (see Fig. 26 for structure) should be the product rather than the *meso*-tartaric acid which is actually found. The experimental data strongly suggested a *trans* addition of bromine. The stereochemistry of the reaction was not determined until 1912 when Alexander McKenzie succeeded in the chemical resolution of the 'isodibromosuccinic acid', thereby establishing it as the (±)-acid, and the 'dibromosuccinic acid' as the *meso* form. McKenzie was unable to provide any rationalization for the *trans* addition of bromine. suggesting only that the transformation bore

> . . . a close analogy to those involving a Walden inversion [see Chapter 7] : the other alternative was that the direct hydroxylation of fumaric acid and maleic acid with permanganate followed a stereochemical course of a different order from that followed by the direct addition of bromine.

Thus it still was not possible to decide which step involved a *trans* addition or substitution. Even the introduction of electronic mechanistic explanations in the 1920s did not lead to a satisfactory understanding of the stereochemistry of the reactions. In 1925, Ethel Terry and Lillian Eichelberger at the University of Chicago thought that the difference in the stereochemistry involved in the reaction of halogens compared with potassium permanganate might be accounted for by the fact that in the basic conditions required for the latter reagent, maleic acid and fumaric acid existed as dianions. They found that, in contrast with the non-ionized

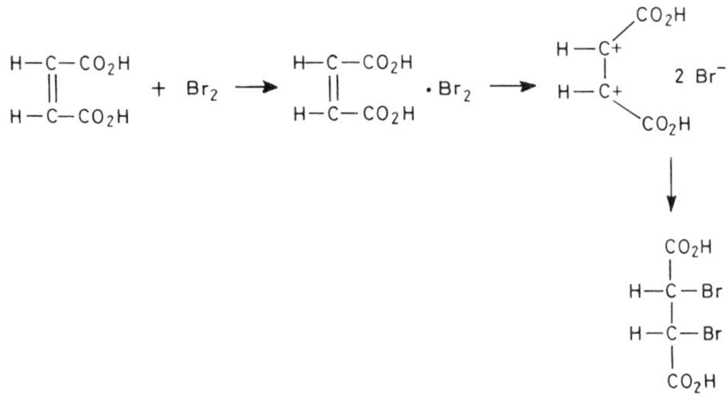

Figure 29. Mechanism proposed by E. Terry and L. Eichelberger for the bromination of maleic acid.

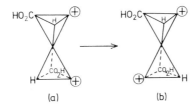

Figure 30. Suggested geometry of dication intermediate.

acids which reacted with Br_2 by a *trans* addition, the dianions reacted to give products that indicated a *cis* addition. The mechanism proposed for the *trans* addition involved the formation of a dipositively charged intermediate (Fig. 29). It was argued that both positively charged carbons retained their tetrahedral geometry (Fig. 30a). The normal *trans* addition was explained in terms of a transfer of a hydrogen atom (the 'lighter of the two groups') with its electrons so that the repulsions between the positive charges were minimized (Fig. 30b). Although a rotation about the carbon–carbon bond could produce the same result (but not the same stereochemistry), it was thought that this would be relatively slow. The *cis* addition in the reaction of the dianions of maleic or fumaric acid was explained by the ability of the negatively charged carboxyl groups to neutralize these positive charges in the intermediate (presumably through the formation of an α-lactone) and thus reduce the tendency for rearrangement.

It was also noted, without comment, that the amount of *cis* addition increased to close to 100% in the chlorination compared to bromination (78%). When the bromination was carried out in an aqueous sodium chloride solution, Terry and Eichelberger reported the formation of two 'mixed' addition products, HOOCCH(Br)CH(OH)COOH and HOOCCH(Br)CH(Cl)COOH, but not the 2,3-dichloro derivative. Inexplicably they interpreted their observations as indicating that the dication intermediate reacted too rapidly to allow it time to trap added ions — although clearly the two 'mixed' addition products contradict this idea. In the same year A. W. Francis obtained similar results when ethylene was brominated in aqueous sodium chloride or sodium nitrate solutions. In these reactions a monocationic intermediate $[BrCH_2 CH_2{}^+]$ was proposed to account for the products formed $[BrCH_2 CH_2 Cl,$ $BrCH_2 CH_2 NO_3]$ under these conditions, but the stereochemistry that might be involved in the reaction was not considered. Ten years later Paul Bartlett and Stanley Tarbell at Harvard University published an article in which they proposed that the effect of added bromide ions on the rate of bromination of stilbene in methanol was consistent with a two-step mechanism involving an intermediate carbonium ion. Although

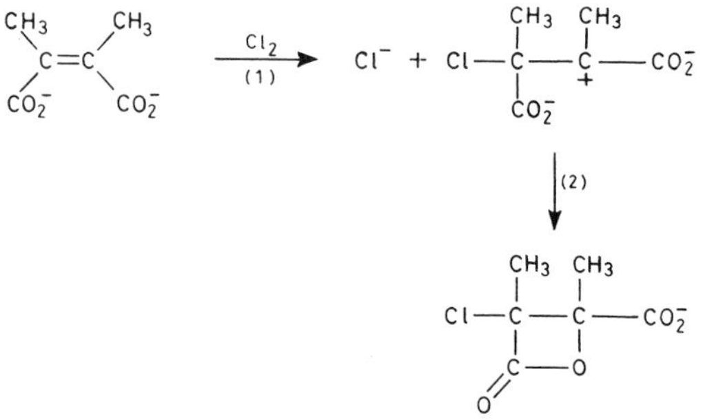

Figure 31.   Proposed mechanism of chlorination of dimethylmaleate
(P. Bartlett and S. Tarbell, 1936).

the stereochemistry of the reaction was not considered in this paper, a year later they were able to demonstrate the stereochemical involvement of the carboxylate ion in the second step of the chlorination of the 2,3-dimethylmaleate anion (Fig. 31).

The stereochemistry involved in the formation of the β-lactone (which was isolated) was explained by assuming that the two steps took place in such rapid succession that no time was allowed for rotation about the carbon—carbon single bond in the intermediate. The fact that addition reactions were stereospecific even in the absence of neighboring anion groups suggested that the trivalent carbonium ion could retain its tetrahedral geometry when it was an intermediate in a reaction (see Chapter 11), even though this was later shown not to be the case.

In 1937, Irving Roberts and George Kimball at Columbia University suggested that the intermediate carbonium ion might have a special stereochemical stability:

If this structure A (Fig. 32) is assumed, one of the orbitals of the C^+ must be completely empty. The X atom on the other hand has three orbitals occupied by pairs of electrons. The arrangement is such that a coordinate link will almost certainly be formed by the sharing of one of the pairs of electrons of the halogen with the unoccupied orbital of the carbon. Another possible structure of the ion is one in which the positive charge is on the halogen. The X^+, being isoelectronic with a member of the oxygen family, should show a valence of two, i.e. it should form a structure of the ethylene oxide type (B).

The *trans* addition products arose via a backside attack by the anion, a process analogous to the Walden inversion (Chapter 7). To account for the *cis* addition:

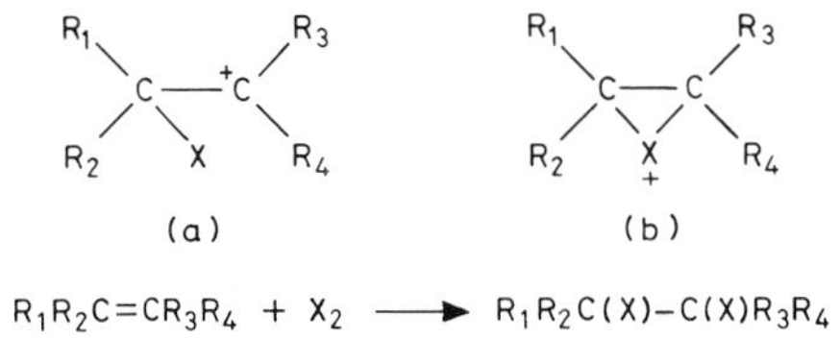

$$R_1R_2C=CR_3R_4 + X_2 \longrightarrow R_1R_2C(X)-C(X)R_3R_4$$

Figure 32. Structure of intermediate cation involved in the halogenation of alkenes.

If, however, R_1 and R_3 are similarly charged groups (e.g., COO^-) there may be sufficient repulsion between them to overcome the restraining force of the double linkage, and rotation to the opposite configuration may take place before the second step of the reaction occurs.

It is more likely that lactone formation accounts for the *cis* addition since this would account for exclusive *cis* addition in chlorination, but less in bromination where the bromonium ion ($X = Br$) produces a more stable intermediate (B).

The research involved in elucidating the stereochemistry of the elimination reaction has a similar but more complicated history, and will therefore not be discussed at any great length. After reviewing a variety of studies at the turn of the century, P. F. Frankland in his Presidential address to the Chemical Society of London concluded that *trans* elimination was far more common than *cis* elimination. He attributed this to the intervention of 'chemical forces' which favored this configuration prior to reaction. By the 1920s, as a result of the mechanistic studies of E. Hughes and C. K. Ingold, it was also clear that the stereochemistry of the reaction was associated with a bimolecular process ('E_2'). In 1939, Saul Winstein, David Pressman and William Young at the California Institute of Technology and the University of California in Los Angeles, reported on the results of an elegant series of studies designed to illustrate the stereochemistry involved in the iodide-catalyzed elimination of diastereoisomeric 2,3-dibromobutanes:

$$CH_3CH(Br)CH(Br)CH_3 + I^{-1} \longrightarrow CH_3CH=CHCH_3 + IBr + Br^{-1}$$

The formation of *trans*- and *cis*-2-butene in high yields from *meso*- and (±)-2,3-dibromobutane, respectively, indicates the strong preference for *trans* elimination in acyclic compounds where there was a good deal of conformational mobility. The process was considered to be analogous to a Walden inversion in which the 'iodide removes a positive bromine atom and essentially simultaneously with this removal the electron pair left unshared by this removal attacks the carbon face opposite the remaining bromine atom forming a double bond and liberating bromide ion'. It is interesting to note, with reference to the later development of conformational analysis (Chapter 14), that the approximately two-fold differences in the rate of elimination of the diastereomers was considered only briefly. The *trans* geometry of the two bromines might be the 'normal state of affairs in dihalides [as revealed by dipole moment measurements] or it may tend to be assumed on the approach by the iodide ion', but how this might affect the rate was not discussed.

7

The stereochemistry of substitution reactions

Plate 12.   Paul Walden (1863–1958) (Edgar Fahs Smith Collection).

In the 1890s the German chemist Paul Walden (1863–1958) (Plate 12) was attempting to ascertain experimentally the relationship between the nature of the groups bound to an asymmetric carbon atom and the magnitude and sign of optical rotation. These studies led to his report, in 1896, of a cycle of reactions which established a chemical relationship between the enantiomers of malic acid:

$$(-)\text{-Malic acid} \underset{KOH}{\overset{PCl_5}{\rightleftharpoons}} (+)\text{-Chlorosuccinic acid}$$

$$\uparrow Ag_2O \qquad\qquad\qquad\qquad \downarrow Ag_2O$$

$$(-)\text{-Chlorosuccinic} \underset{KOH}{\overset{PCl_5}{\rightleftharpoons}} (+)\text{-Malic acid}$$
$$\text{acid}$$

It was clear that an inversion of configuration at the asymmetric carbon atom had taken place; it was not clear, however, which of the reagents was responsible for the inversion. Walden's observations were a great puzzle to chemists, since up to that time it was assumed that any substitution should proceed with retention of configuration (that is, the new substituent would simply replace the original without any molecular rearrangement) if it led to an optically active product, inasmuch as the other experimental conditions known at that time that involved an inversion produced a racemic product. As an illustration of the pervasiveness of this view (even after Walden's report), P. F. Frankland stated (in 1913) that it was assumed that the group should either be replaced with retention or '. . . that it should take up any other position *without* preference . . .' to give a racemic mixture.

Walden thought that PCl_5, as well as strong bases, reacted by the 'normal' substitution (that is, with retention of configuration), in contrast to the heavy metal oxides which produced an 'abnormal' substitution (leading to an inversion of configuration).

To explain the 'abnormal' reaction, Walden proposed that the carboxyl group was involved in the formation of a lactone with inversion of configuration. The second step, which led to the product, was assumed to proceed with retention of configuration:

$$\begin{array}{c} ClCHCOOAg \\ | \\ CH_2-CO_2H \end{array} \xrightarrow{①} \left[\begin{array}{c} \overset{O}{\overset{\diagup\diagdown}{CH-CO}} \\ | \\ CH_2-CO_2H \end{array} \quad or \quad \begin{array}{c} -CH-COOH \\ | \\ CH_2-CO \\ -O- \end{array} \right] \xrightarrow[②]{H_2O} \begin{array}{c} HOCH-COOH \\ | \\ CH_2-COOH \end{array}$$

Walden's mechanistic proposals do not seem to have been considered seriously by his contemporaries; only later was Walden's name associated with the stereochemistry of the inversion step.

The term 'Walden inversion' was first used by Emil Fischer in 1906 to designate a series of reactions in which an optically active substance was converted to its enantiomer. By the turn of the century, examples of the Walden inversion had become so numerous that many chemists had begun to suspect that those reactions involving a retention in configuration might have to be considered 'abnormal'. The explanations proposed to

Figure 33. Model used by Emil Fischer to illustrate the stereochemistry
of a Walden inversion reaction.

account for the inversion, however, were generally quite unsatisfactory
since they were unable to predict what conditions or reagents might
produce the inversion. Of these early theories, those proposed by Emil
Fischer and Alfred Werner were most favorably received. In Fischer's
view, the 'residual affinity' not used in the bonding of the four groups to
an asymmetric carbon, could be used in the formation of an 'addition-
complex' with the displacing reagent. Following the separation of the
halogen, in an ionized form, the displacing reagent could then either (1)
take the place of the halogen (which results in retention of configuration),
or (2) one of the other groups already bound to the carbon could take its
place, followed by the reagent taking up this group's original position
(which would produce the inversion). The process was illustrated by a
model, in which the asymmetric carbon was covered with a bristle, to
show how all five groups could be joined to form the addition complex
(Fig. 33). The model could be used to indicate the stereochemistry of the
reaction of 2-bromopropanoic acid with ammonia:

$$\overset{*}{CH_3}CH(Br)COO^{-1}NH_4{}^{+1} + NH_3 \rightarrow \overset{*}{CH_3}-CH(NH_2)COO^{-1}NH_4{}^{+1} + HBr$$

Spheres 1−4 represent H, Br, CH_3 and $COO^-NH_4{}^+$ bonded to the asymmetric carbon, 7, which is shown in its pentavalent 'addition-complex'.
Atoms 5 and 6 are two of the hydrogens bound to nitrogen.

Werner's idea of 'primary' and 'secondary' valencies, used so success-
fully to explain the formation of inorganic complexes, was extended to
carbon compounds. In the displacement reaction, an intermediate complex was formed through the operation of the attractive forces in the

secondary sphere of carbon. The first sphere was then penetrated by the reagent while the halogen simultaneously became part of the second sphere. Whether this resulted in an inversion or retention was determined by the position the reagent occupied in the first sphere, and this was controlled by the direction of the net attractive force emanating from the asymmetric carbon atom into the second sphere.

The deficiencies of this theory are apparent. Not only was it not possible to decide which reactions might produce an inversion product, but also there was no experimental evidence available that could be cited to support the existence of such an intermediate complex.

By the turn of the century the idea of tetrahedrally oriented valence-forces in carbon compounds had become so well established, that proposals such as Werner's were looked upon as a reversion to earlier views of chemical affinity which acted 'uniformly from a center of the assumed spherically shaped atom', to quote from a critical commentary in the *Annual Reports of the Chemical Society* in 1912. The theories proposed by other chemists (the names of Pfeiffer, Gadamer and Meisenheimer might be mentioned) in this period were similar to those of Fischer and Werner but were no more successful in providing a deeper insight into the mechanics of the inversion. Experimental observations continued to be reported that could not be accommodated by the theories. For example, E. Biilman pointed out (in 1911) that Fischer's requirement that (in the reaction of ammonia with an alkyl halide) one of the ammonia hydrogens had to split off before the displacement occurred was inconsistent with displacement involving triethylamine. Biilman proposed that all displacement reactions involved an intermediate carbon cation in which the asymmetry of the original carbon was maintained. That the carbocation might maintain the same geometry as its tetrahedral precursor was not considered theoretically impossible. For example, Frankland noted:

> that an ionized quadrivalent atom should be capable of preserving its asymmetry is not so surprising, if ionization be regarded as resulting from the combination of the electrolyte with the maximum number of water molecules which it can hold by means of the residual valency of the oxygen.

The reports of optically stable compounds containing positively charged heteroatoms, such as the trialkyl sulfonium halides, in the same period (Chapter 10) provided additional support for the reasonableness of a view that the carbon ion could also exhibit stereochemical stability. The complete understanding of the stereochemistry involved in the Walden inversion continued to be delayed by a number of experimental problems, four of which are discussed briefly below:

1. The compounds initially used in the studies of the Walden inversion were generally α-halo substituted carboxylic acids or dicarboxylic acids. As these compounds were studied more systematically it was found that variations in the concentrations of the base used as displacing agents gave products of varying optical purity. The cause of these variations could not be understood until the 1930s when, as a result of kinetic studies, it was shown that α-halocarboxylic acids reacted in basic solution by both a first-order and second-order process. The former gave a product with a retention of configuration, while the latter involved an inversion. In 1937, Cowdy, Hughes and Ingold showed that (+)-2-bromopropanoic acid in 1 M sodium methoxide in methanol underwent a second-order reaction to give (−)-2-methoxypropanoic acid. In 0.1 M sodium methoxide, the reaction was first order and produced the (+)-enantiomer.

2. The extensive studies of G. Senter at Birkbeck College in London from about 1915 to 1925 demonstrated that whether inversion or retention of configuration is observed could be controlled by the solvent used. For example, in the reaction of (−)-α-bromophenylacetic acid with ammonia, when water or ethanol was used as the solvent, (+)-α-amino-phenylacetic acid was the product; when the solvent was changed to acetonitrile or liquid ammonia, the (−)-enantiomer was formed. (In other studies it was also shown that the stereochemistry of the reactions could be significantly affected by the temperature of the reacting system.)

3. The nature, and concentration, of the metal ion associated with the displacing base was also shown to affect the optical purity of the product. Senter observed that the optical purity of the product was not only affected by the concentrations of the heavy metal present (such as silver ions) but also by the amount of the precipitated silver halide formed.

4. It was also not clear whether the processes that led to a racemic product were necessarily the same as those involved in the Walden inversion. In 1904, for example, Werner had proposed that racemization involved the movement of the four groups about the tetrahedral carbon into a coplanar configuration.

Finally, as had been suggested earlier, it still was not clear how one would go about establishing unambiguously which reagent and conditions in a series of reactions were responsible for the inversion. This was not accomplished until the 1920s by H. Phillips, who conceived that this could be determined by studying a series of reactions in which only one step involved the cleavage of a bond to the asymmetric carbon atom. The reactions studied by Phillips are summarized in Fig. 34. Since steps (1) and (2) do not involve the breaking of a bond to the asymmetric carbon, the inversion must take place in either step (3) or (4). But since the reaction of the (+)-ester with hydroxide in step (1) regenerates the original alcohol, this means that step (4) cannot also be responsible for an inversion of configuration. Step (3) therefore is responsible for the inversion.

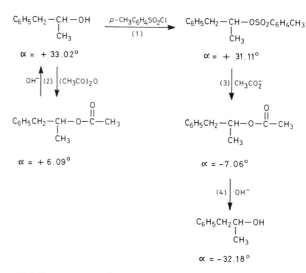

Figure 34. Reactions studied by Phillips to establish the identity of the inversion step.

Although the mechanism proposed by Phillips to explain the inversion was still no more satisfactory than those put forth earlier, his experimental approach did provide chemists with a method by which they could determine which reagents were responsible for inversion.

That the inversion involved a synchronous backside attack of the displacing reagent as the displaced group was ejected was not generally accepted by chemists until the 1930s. Le Bel had proposed such an explanation in 1911, but only in rather general terms to explain the effect of the reactant on the 'stable' equilibrium of the four groups attached to an asymmetric carbon. In 1923, G. N. Lewis (1875–1946) argued that a backside attack was the most reasonable means by which an inversion could be explained:

> There seems to be but one possible way of accounting for this peculiar behavior [i.e. of the Walden inversion]. Let us consider a carbon atom attached to the four radicals R_1, R_2, R_3 and R_4 and let us assume that a fifth group, R_5, becomes temporarily attached to the carbon atom near the face of the tetrahedron which is opposite R_1. A slight shift of the kernel might make it now the center of a new tetrahedron with corners at R_2, R_3, R_4 and R_5, while R_1 would become detached from the molecule; then if the radical R_5 in the new molecule were to be replaced by the radical R_1, the resulting molecule would be the

mirror image of the one with which we started. In this explanation it is not necessary to assume that the five radicals are attached to the carbon for any appreciable period of time, indeed it might be assumed that the R_1 leaves at the same instant that R_5 becomes attached to the carbon atom. . .

Although such arguments could be used to explain reactions involving an inversion, it was by no means clear how the formation of racemic products was to be explained. The answer awaited the development of physical organic chemistry. In the 1930s, E. D. Hughes and C. K. Ingold demonstrated the relationship between the kinetic behavior of substitution reactions and their stereochemistry. When the 'molecularity' was correlated with the stereochemistry of a reaction, it was found that bimolecular nucleophilic substitution ('S_N2') led to an inversion of configuration; whereas in a unimolecular process ('S_N1') a racemic product was generally produced. In the latter case, the mechanism involved the attack of the nucleophile on either side of the planar carbonium ion that was formed as an intermediate.

Persuasive experimental evidence in support of the proposal that every substitution in a bimolecular process is accompanied by inversion was furnished by Hughes in 1935 in the reaction of (+)-2-iodoctane with radioactive iodide:

$$(+)\text{-}C_6H_{13}CH(I)CH_3 + \overset{*}{I}{}^- \longrightarrow (-)\text{-}C_6H_{13}CH(\overset{*}{I})CH_3 + I^-$$

where the rate of exchange was found to equal the rate of inversion (that is, it was one-half the rate of racemization). As the stereochemistry of the inversion process became better understood, questions began to be asked about the kind of bonding that might be involved in a transitional state containing both the displacing and displaced group. T. M. Lowry had suggested (in 1925) the involvement of an intermediate carbonium ion in all displacement reactions, but since he was unable to make any *a priori* predictions as to whether a reaction would proceed by inversion, retention or racemization, the proposal did not achieve any degree of popularity.

In 1932, N. Meer and M. Polanyi argued that the bond dipole of the 'anionic' group to be displaced would help direct the approach of negative ions (e.g. OH^-) to the side away from the group, which allowed them to distinguish between a 'negative mechanism' of the type that produced inversion, and a 'positive mechanism' involving a cation, which presumably led to a retention of configuration. It was soon pointed out, however, that the 'anionic mechanism' could not explain the stereochemistry of the reactions of anions with quaternary ammonium salts, where the positive charge on nitrogen would favor a front-side attack. For example:

$$OH^- + CH_3N(CH_3)_3{}^{+1} \longrightarrow HOCH_3 + N(CH_3)_3$$

In 1933, A. R. Olson provided an explanation of the bimolecular displacement in terms of the quantum mechanical ideas recently proposed by Linus Pauling and R. H. Slater. When considering the carbon–chlorine bond in a chloroalkane, for example, Pauling and Slater had noted that the atomic orbital of carbon extended 'beyond the carbon atom on the side away from the chlorine'. The carbon atom therefore has the possibility of bond formation at the face of the tetrahedron opposite the chlorine atom:

> The group which is to be displaced determines a unique path for the entering group, such that the system requires less energy than it would for any other path. This, then, leads to complete inversion of configuration for a reaction which takes place in one step.

Where backside attack is prevented, nucleophilic substitution reactions are effectively blocked. Paul Bartlett and L. H. Knox observed in 1939 that 1-chloroapocamphene (Fig. 35a) was completely inert to a boiling solution of sodium ethoxide in ethanol (as well as to an ethanolic solution of $AgNO_3$), whereas the corresponding acyclic compound (Fig. 35b) was very reactive.

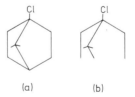

(a)   (b)

Figure 35.

In this same period, a number of inorganic chemists were investigating the stereochemistry of displacement reactions of inorganic complexes. No series of reactions of optically active inorganic compounds, analogous to those discovered by Walden, were known until John Bailar's studies in

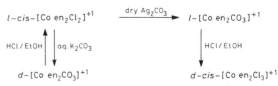

Figure 36. Interconversions of enantiomers of some cobalt complexes
(Bailar, 1934).

1934 involving the reaction of cobalt complexes (Fig. 36). Although these reactions were at first described as examples of 'inorganic Walden inversions', it could not be determined which reagent (K_2CO_3 or Ag_2CO_3) produced the inversions taking place in organic compounds. (More details of the stereochemistry of inorganic displacement reactions can be found in Kauffman's *Inorganic Chemical Complexes* in this series.)

8

Symmetry, asymmetry, chirality

Although the terms 'asymmetry' and 'asymmetric carbon' have been widely used by organic chemists, it has only been in recent years that attention has been given to the proper meaning of the terms. In a paper published in 1966, R. S. Cahn, C. K. Ingold and V. Prelog indicated the changes that should be made to clarify the language used in stereochemical discussions:

> Chemists habitually use the word 'asymmetry' (and its grammatical congeners) in two different senses, sometimes without appreciating the differences. The symmetry of any molecular model consists in the totality of independent symmetry operations of (a) rotation (by sub-multiples of 2π) round the axes of symmetry, (b) reflexion in planes of symmetry, and (c) combinations of such a rotation and a reflexion, which will bring the molecule into coincidence with itself. Only when no symmetry operation can so convert the molecule, has the molecule no symmetry. Only then may it be called 'asymmetric' in the correct usage of the term.
>
> Asymmetry is a sufficient condition for the existence of optical enantiomers, but it is not, though it is sometimes referred to as if it was, the necessary and sufficient condition. The necessary and sufficient condition is that reflexion in a plane converts the model into a non-identical one, that is, one which cannot be superposed on the original by translation and rotations only. The model then has two non-identical forms, interrelated by a reflexion, that is, two enantiomeric forms: it has the topological property of handedness.

To avoid further confusion they suggested the use of the word 'chirality' to designate the 'handedness' of a molecule, and to use the term 'asymmetry' only as defined by the symmetry operations described above. Apparently the word *chiral* was first used in 1893 and 1904 by Lord Kelvin, and by L. L. Whyte in 1957–58, but has only been widely adopted by organic chemists in recent decades. In the 1920s and 1930s, F. M. Jaeger at the University of Groningen had suggested that the term *dissymmetry* be used to indicate 'handedness':

> [Pasteur] . . . emphasized the fact that the one configuration [of the optically active molecules] was the non-superposable mirror image of the other owing to the absence of some symmetry properties in their arrangement, or as he expressed it, the existence there of a certain 'dissymmetry'. According to Pasteur it is, therefore, not at all the completely asymmetric molecules alone that will show this isomerism.

Unfortunately, in the published English version of Pasteur's 1860 lecture, the word *dissymétrie* is translated as asymmetry. This mistranslation, along with Van't Hoff's emphasis on the importance of the *asymmetric* carbon atom, probably accounts for the continued use of the word asymmetry instead of *dissymmetry*. In the 1940s, George Wheland also urged that the term *dissymmetry* be used to describe any molecule that lacks an alternating axis of symmetry.

To summarize the current view: a molecule that has a plane, center, or an alternating axis of symmetry will be found to be superimposable on its mirror image and therefore will be optically inactive (*achiral*). A molecule that has no element of symmetry (an asymmetric molecule) or one that has only a simple axis of symmetry, is not superimposable on its mirror image and will be optically active (*chiral*). Thus the conformations of molecules that exhibit no reflection symmetry are referred to as *dissymmetric* or *chiral*, the latter term being preferred. (See the Glossary for further discussion.)

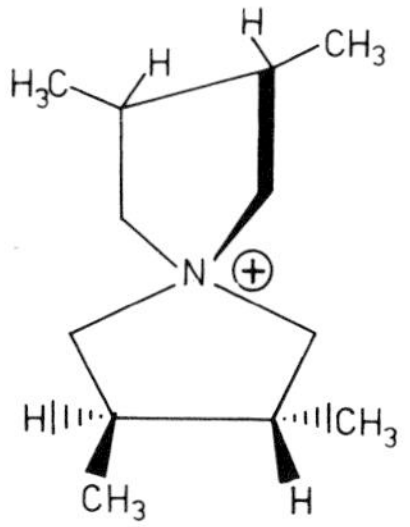

Figure 37.

It was not until the 1960s that chemists began to appreciate the importance of the symmetry properties of molecules to the understanding of a number of stereochemical problems. Earlier most chemists were satisfied that, for example, an optically inactive molecule contained a plane or center of symmetry. That this was not the only condition, however, had been pointed out in 1903 by E. Mohr (and by O. Aschan even earlier) who observed that while molecules of the type $Z(A^+A^+A^-A^-)$ may lack a plane or center of symmetry, they are nevertheless image-superimposable. The synthesis of the first example of this type of compound (Fig. 37) was reported by J. E. McCasland and S. Proskow in 1956. The meso-*trans*/ *trans* diastereomer of 3,4,3',4'-tetramethylspiro-(1,1')-bipyrrolidinium *p*-toluenesulfonate is optically inactive even though it has no plane or center of symmetry. (It does, however, possess a four-fold alternating axis of symmetry.)

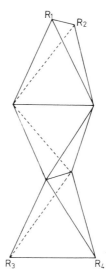

Figure 38. Asymmetric allene, $R_1 R_2 C=C-CR_3 R_4$ (Van't Hoff, 1875).

Most of the chiral compounds encountered in stereochemical research contain at least one asymmetric carbon. In 1875, Van't Hoff provided an example of a molecule that would be expected to be optically active, even though the molecule did not contain any asymmetric carbon atoms (Fig. 38). For allenes of the general formula $R_1 R_2 C = C = CR_3 R_4$, two non-superimposable mirror-image isomers were possible since the R_1, R_2 groups were not found in the same plane as R_3 and R_4. Later it was realized that this condition was fulfilled even for simpler allenic compounds (for, e.g., $R_1 R_2 C = C = CR_1 R_2$). The relationship of the allene

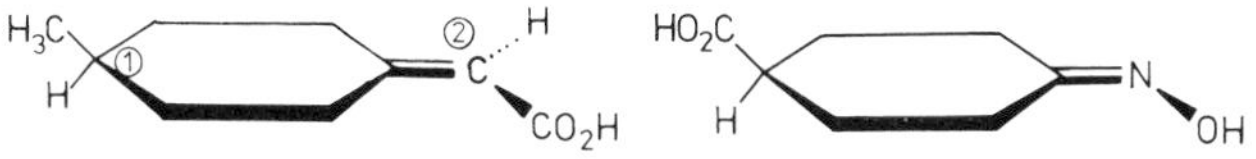

(a)   (b)

Figure 39.

structure to the tetrahedron can be more readily visualized by considering that the four substituents are located at the corners of an elongated tetrahedron.

The first resolution of a compound (Fig. 39a) structurally similar to the allenes was reported in 1909 in a joint publication of William Perkin, William Jackson Pope and Otto Wallach (at the Universities of Manchester, Cambridge and Göttingen, respectively). Shortly thereafter W. H. Mills and Alice M. Bain, at the Northern Polytechnic Institute in London, were able to obtain an optically active sample of the oxime of 4-carboxycyclohexanone (Fig. 39b). The isolation of this compound not only provided an additional example of an optically active compound which did not contain an asymmetric carbon atom, but also provided significant support for the Hantzch–Werner hypothesis concerning the stereochemistry of oximes (Chapter 11). Some chemists argued that carbon number 1 (Fig. 39a) should be considered an asymmetric carbon since the configuration of the molecule was not the same when taken either way around the ring. The resulting controversy forced chemists to reconsider what was meant by the term 'asymmetric carbon atom'. The chemist John Marsh, for example, argued that if the exchange of any two of the groups about a carbon brings about a change in the formula as a whole (that is, to produce its enantiomer *or* a geometrical isomer), then that carbon should be considered asymmetric. However, if this argument is allowed, it is found that not only does carbon number 1 qualify as an asymmetric carbon atom but also carbon number 2. With the acceptance of the Cahn–Ingold–Prelog method of assigning configurations (Chapter 9), a solution to this problem was available. In this compound, neither carbon can be considered an asymmetric center since it is not possible to make an R/S configurational assignment. (It is interesting to note, however, that the molecule does possess a *chiral axis*.)

The first preparation of an optically active allene (Fig. 40) was reported in 1935 by W. H. Mills and Peter Maitland at Cambridge. The report was also interesting since the synthesis involved an asymmetric dehydration using optically active *d*-camphorosulfonic acid as a catalyst (see Chapter 15). The specific rotation (at 5461 Å) of the product was found to be +437° when (+)-camphorosulfonic acid was used, and −438° when the (−)-enantiomer was used. (This, as we now know, corresponds to the presence of about 53% of one enantiomer.)

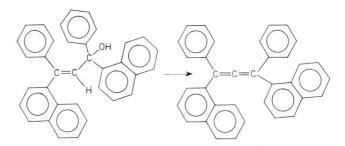

Figure 40.   Synthesis of an optically active allene (Mills and Maitland, 1935).

The preparation of an optically active allene by a conventional resolution was reported in the same year by E. P. Kohler, J. T. Walker and M. Tishler at Harvard University by recrystallizing the brucine salt of a derivative of 1,3-diphenyl-3-naphthylallene carboxylic acid, $(C_{10}H_7)(C_6H_5)C = C = C(C_6H_5)COOCH_2COOH$.

Another class of chiral compounds that do not contain an asymmetric carbon (chiral center) are the spiranes. The term spirane (Latin *spira*, twisted) was coined in 1900 by Adolf Baeyer, who considered the geometry of the compounds analogous to a pretzel. (It has been pointed out that the two rings in a spirane are perpendicular to each other, unlike the rings in a pretzel which are coplanar.) Werner and Aschan later pointed out that molecules with this geometry could be optically active. The first spiro compounds that were resolved were heterocyclic compounds — generally substances containing a nitrogen atom as the central atom that joined the two rings. The first report of the resolution of a completely carbocyclic spirane (Fig. 41a) was given by S. E. Janson and W. J. Pope in 1932. (A partial resolution of another spiro compound (Fig. 41b) had been reported a year earlier by Backer and Schurink.)

The numerous reports made in the 1920s and 1930s concerning the isolation of optically active allenes, spiro compounds and related substances suggests that chemists had settled into a program of research that did little more than provide additional examples of unusual chiral molecules that did not contain asymmetric centers.

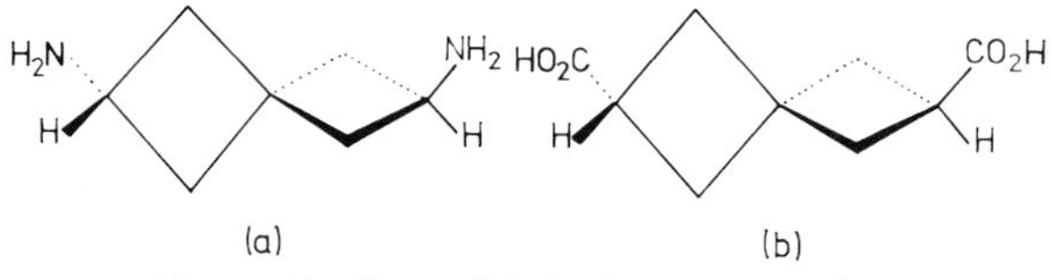

Figure 41. Some chiral spiro compounds.

The availability of isotopes for synthetic uses in the 1930s stimulated research directed toward determining whether the difference between an element and its isotope was sufficient to contribute to a molecule's chirality. The early studies were ambiguous, since it was not possible to rule out the presence of impurities that might have produced the low values of optical rotation. The first unambiguous preparations of compounds whose chirality depended entirely on isotopic substitution (deuterium) was reported in 1949 by E. R. Alexander and A. G. Pinkus, and E. L. Eliel. Eliel's synthesis of optically active α-deutero-ethylbenzene via a Walden inversion reaction using lithium aluminum deuteride is shown in Fig. 42.

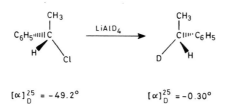

$$[\alpha]_D^{25} = -49.2°$$ $$[\alpha]_D^{25} = -0.30°$$

Figure 42.

9

The specification of molecular configuration

It was because chemists such as Johannes Wislicenus (Chapter 6), Adolf von Baeyer (Chapter 13) and Emil Fischer were able to demonstrate the practical applications of Van't Hoff's ideas that the basic concepts of stereochemistry were widely accepted by the end of the 19th century. As chemists in new fields adopted these concepts it also became clear that a systematic basis was needed for the specification of molecular configuration. It is to Emil Fischer that we must give the major credit for the establishment of such a system, and this he did as a result of his researches on carbohydrates. That Fischer was able to bring order to such a complex field attests to his genius. It would be well worth the reader's time to consult the historical accounts of his work.

Problems in the assignment of configuration cannot be easily solved using pencil and paper; the relationship between stereoisomers can best be seen with the use of molecular models. As Fischer recalled in his auto-biography:

> I remember especially a stereochemical problem. During the winter of 1890–91 I was busy with the elucidation of the configuration of sugars but I was not successful. Next spring in Bordighera [where Fischer was accompanied by Adolf von Baeyer] I had an idea that I might solve the problem by establishing the relation of pentoses to trihydroxyglutaric acids. However, I was not able to find out how many of these acids are possible; so I asked Baeyer. He attacked such problems with great zeal and immediately constructed carbon atom models from bread crumbs and toothpicks. After many trials he gave up because the problem was seemingly too hard for him. Only in Würzburg by long and careful inspection of good models did I succeed in finding the final solution.

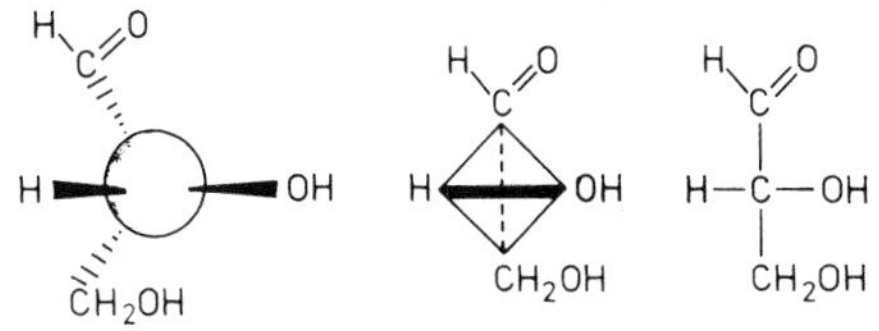

Figure 43.   Perspective and Fischer projection formulas of (+)-glycer-
aldehyde.

As a result of extensive systematic studies of a series of carbohydrates,
Fischer was able to determine the stereochemical relationship between a
large number of sugars. Originally, the configurations of all of the isomeric
and stereoisomeric sugars were related to an arbitrarily assigned configuration of the sugar acid, saccharic acid. (Later (+)-tartaric acid was used;
finally (+)-glyceraldehyde.) Fischer proposed the use of a two-dimensional
drawing to illustrate the configurations of the sugars based on molecular
models used to show the configurational orientation of the four groups
about asymmetric carbon atoms. The model of the sugar molecule was
turned in such a way that the main carbon chain containing the aldehyde
and hydroxymethyl group could be visualized as lying beneath a plane (of
the paper), and the other groups about the asymmetric carbon atoms,
above the plane. Two ways in which (+)-glyceraldehyde might be represented in three dimensions are shown in the first two figures in Fig. 43.
The two-dimensional Fischer projection formula (third figure) is produced
by pressing the four groups into the same plane. This configuration of
(+)-glyceraldehyde was designated by chemists for many years as d-glycer-
aldehyde. The use of the uncapitalized letters d and l accounts for some of
the confusion in the next 50 years since the letters were also used to indicate whether a substance was dextrorotatory or levorotatory. The configurational assignment was later indicated by the capitalized letters D and
L. Even this change did not seem to clarify the situation because the
meaning of D and L was often not well defined. By the 1950s the sign of
optical rotation was denoted by a plus or minus sign in parentheses [(+)
or (−)] and this convention will be used in this discussion.

In Fischer's original work, the configurational assignment of sugars
containing more than one asymmetric carbon was based upon chemical
methods. Although the methods are discussed in some detail in books on
carbohydrate chemistry, a few of the problems that arose from the use of
this method might be mentioned. Most contemporary discussions point
out that the D- or L- assignment for a sugar, such as glucose, is based upon
the orientation of the hydroxyl on the asymmetric carbon furthest removed from the aldehyde carbon. Thus D-glucose has the configuration
shown below:

$$
\begin{array}{ccc}
\overset{\displaystyle O}{\underset{\displaystyle |}{\overset{\displaystyle \|}{CH}}} & \overset{\displaystyle O}{\underset{\displaystyle |}{\overset{\displaystyle \|}{CH}}} & \overset{\displaystyle O}{\underset{\displaystyle |}{\overset{\displaystyle \|}{COH}}}
\end{array}
$$

O ‖ CH	O ‖ CH	O ‖ COH
H—C—OH	HO—C—H	H—C—OH
HO—C—H	HO—C—H	HO—C—H
H—C—OH	H—C—OH	H—C—OH
H—C—OH	HO—C—H	H—C—OH
CH₂OH	CH₂OH	COH ‖ O
(+)-Glucose	(−)-Gulose	(+)-Saccharic acid

Since (+)-glucose and (−)-gulose both gave (+)-saccharic acid on oxidation, Fischer assigned both of them to the same *d*- (or later D-) configuration; which in the case of (−)-gulose, is contrary to the convention just mentioned. (Note: In the oxidation of the terminal carbons of gulose to carboxyl groups, the molecule must be rotated 180° in the plane of the paper, to produce (+)-saccharic acid.)

The problem was not appreciated until 1906 when M. S. Rosanoff pointed out that the assignment of the configuration of the two carbohydrates to the same D-series on the basis of the results of a set of chemical reactions, such as Fischer had done, would lead to a number of inconsistencies, in addition to the one already mentioned. It is not possible to discuss Rosanoff's paper in any detail; suffice it to say, he proposed the convention now in use. Under Rosanoff's convention, (−)-gulose was made a member of the L-series. (In an attempt to get away from the confusion produced by the use of the D- and L- prefixes, Rosanoff also proposed the use of the Greek letters δ- and λ-; but his proposal was not adopted by other chemists.)

The assignment of the configuration of (+)-tartaric acid has met with difficulties. From an examination of the formula, it is not possible to decide which asymmetric carbon atom should be used to assign the D- or L- configuration. Fischer had assigned it as a member of the D-series on the basis of its synthesis from (+)-glucose. This basis for the assignment was rejected by Rosanoff with the comment: 'The direct oxidation which produced ordinary tartaric acid from *d*-glucose is scarcely a more reliable criterion than would be a process of destructive distillation.' By this time chemical methods used to assign configuration had also become suspect since the observations of Walden had shown that all replacement reactions could not be depended upon to lead to a retention of configuration (Chapter 7).

In 1914–22, K. Freudenberg assigned (+)-tartaric acid to the D-series on the basis of the synthetic relationship he had established between D-malic acid, D-lactic acid, D-glyceric acid and D-glyceraldehyde (Fig. 44).

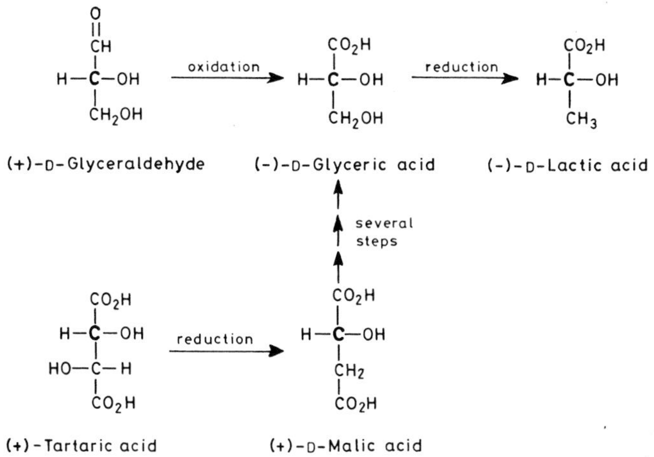

Figure 44.   Synthetic method used by K. Freudenberg (1914–22) to
             assign the configuration of (+)-tartaric acid.

In 1917, A. Wohl assigned (+)-tartaric acid to the L-series based on
another series of reactions that led to the preparation of (−)-tartaric acid
from (+)-D-glyceraldehyde (Fig. 45). (The location of the asymmetric
carbon originally derived from (+)-D-glyceraldehyde is indicated in these
figures by a bold-faced letter.) The fact that these two chemical methods
led to the assignment of opposite configurations for tartaric acid seriously
undermined the validity of configurational assignment by chemical means.
Although attempts have been made to demonstrate the greater reasonableness (or unreasonableness) of one or the other of the methods, no agreement has been reached to date. The reader is warned that both conven-
tions are still used in textbooks, and often the authors do not seem to be
aware of the reasons for the ambiguity.

The D- and L- convention was also used to designate the configuration
of compounds other than carbohydrates, and worked reasonably well for
many compounds. The widespread use of the term 'L-amino acids' attests

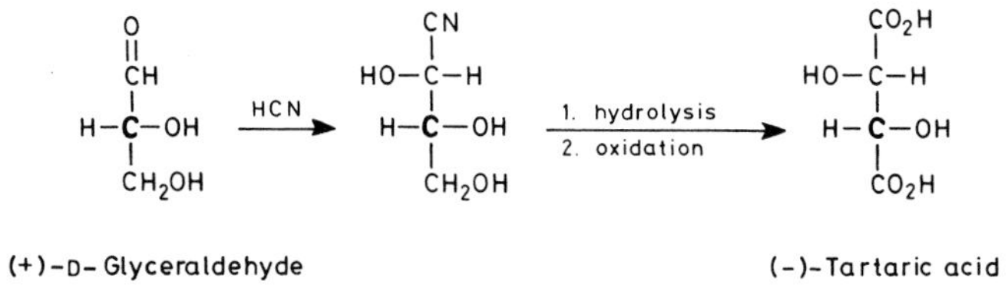

Figure 45.   Synthetic method used by A. Wohl (1917) to assign the
             configuration of (−)-tartaric acid.

to its general acceptance in this field. Thus, for example, (+)-alanine has the same configuration as L-(+)-lactic acid:

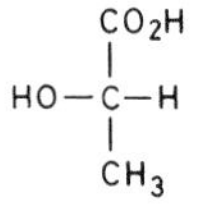

$$
\begin{array}{c}
CO_2H \\
| \\
HO-C-H \\
| \\
CH_3
\end{array}
\qquad
\begin{array}{c}
CO_2H \\
| \\
H_2N-C-H \\
| \\
CH_3
\end{array}
$$

L-(+)-Lactic acid 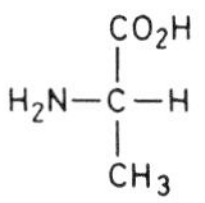   L-(+)-Alanine

The extension also worked reasonably well for compounds such as RCHXR′, where R−C−R′ constituted the main carbon chain and X represented some heteroatom or group containing electronegative elements. Difficulties arose in more highly substituted compounds (RR′CXR″ or RR′CR″R‴). To avoid misunderstanding, it was suggested that subscripts be used to indicate whether the configurational assignment was based on glucose (g) or serine (s). Thus, for example (−)-threonine might be written as either D_g-(−)-threonine or L_s-(−)-threonine:

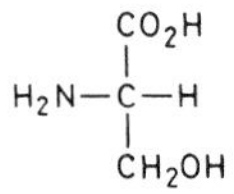

$$
\begin{array}{c}
CO_2H \\
| \\
H_2N-C-H \\
| \\
CH_2OH
\end{array}
\qquad
\begin{array}{c}
CO_2H \\
| \\
H_2N-C-H \\
| \\
H-C-OH \\
| \\
CH_2OH
\end{array}
$$

L-(−)-Serine 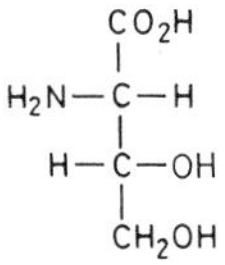   (−)-Threonine

In an attempt to broaden the applicability of the D/L notation, W. Klyne suggested in 1951 that the configuration of each asymmetric carbon atom should be specified. As an illustration, the compound shown below would be named (−)-pentane-2L,3D-diol:

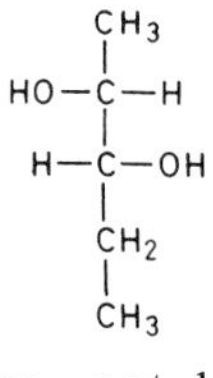

$$
\begin{array}{c}
CH_3 \\
| \\
HO-C-H \\
| \\
H-C-OH \\
| \\
CH_2 \\
| \\
CH_3
\end{array}
$$

A completely new convention, not based on the Fischer projection formulas, was proposed by R. S. Cahn, C. K. Ingold and V. Prelog in the 1950s: To designate the configuration about any particular carbon, the four groups (a, b, c, d in Fig. 46) on the carbon are arranged according to a sequence based on the order of decreasing atomic number. The molecule is then viewed so that the group having the lowest sequence number (d) is the most remote from the viewer. If the remaining three groups trace (a → b → c) a counter-clockwise turn, the configuration about that carbon is designated as S (Latin *sinister*, left); if a clockwise turn, by R (Latin

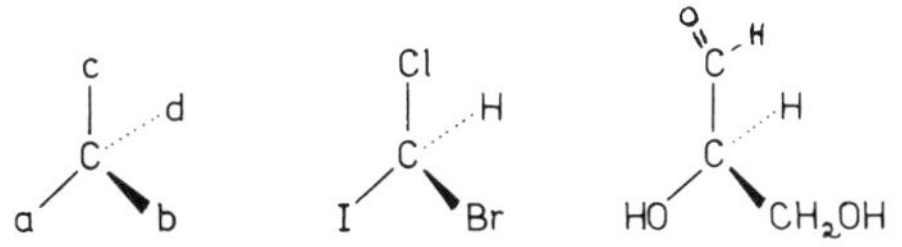

(S)-Bromochloroiodomethane
(R)-(+)-Glyceraldehyde

Figure 46. Application of the Cahn–Ingold–Prelog method for specifying
molecular configuration.

rectus, right). (The configuration specifications of particular stereoisomers
of bromochloroiodomethane and (+)-glyceraldehyde by this method are
illustrated in Fig. 46.) By this method, an unambiguous configurational
name can be given for (+)-tartaric acid: (R)-(+)-tartaric acid [$(2R,3R)$-
dihydroxysuccinic acid, or $(2R,3R)$-dihydroxybutanedioic acid are also
used].

Extensions of the method allow for the configurational assignment of
chiral compounds that do not contain chiral centres (examples found in
Chapter 8). Allenes and similar compounds can be considered as elongated
tetrahedra through which passes what is designated as a 'chiral axis'. The
configuration is assigned by looking at the disposition of the four groups
about this axis. By looking along the axis, we can distinguish two 'near'
groups and two 'far' groups. Two-dimensional projection formulas of the
compounds (found in Figs. 39a, 39b and 54) are illustrated below:

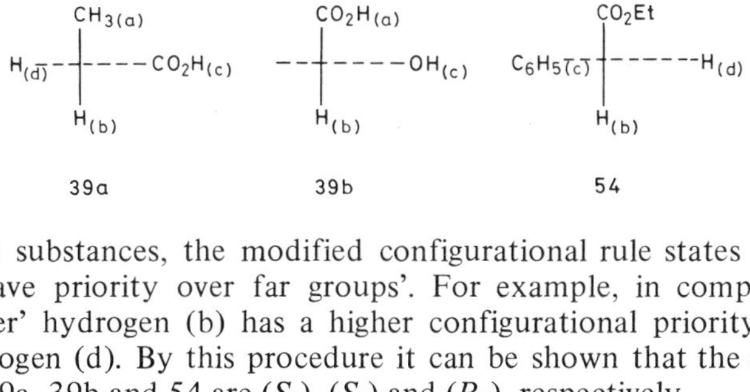

39a   39b   54

For these substances, the modified configurational rule states that 'near
groups have priority over far groups'. For example, in compound 39a
the 'nearer' hydrogen (b) has a higher configurational priority than the
'far' hydrogen (d). By this procedure it can be shown that the configurations of 39a, 39b and 54 are $(S-)$, $(S-)$ and $(R-)$, respectively.

More recently the rule has been modified so that one views the molecule from the 'inside out' rather than 'outside in', thus bringing the cases
of axial chirality in closer agreement with those having chiral centres. The
reader is urged to consult some of the more recent books on stereochemistry for a more detailed discussion.

For *cis/trans* isomers, the terms *seqcis* or Z (German, *zusammen*) and
seqtrans or E (German, *entgegen*), based on the configurational priorities
of the groups bound to the carbons of the carbon–carbon double bond,

has come into greater use. Briefly summarized: one determines which group on each carbon has the higher precedence. The notation is then based on whether the groups having the higher precedence are on the same side (Z) or the opposite side (E). The system avoids confusion but may not provide a correspondence with earlier methods:

cis / trans Nomenclature: cis-1,2-dichlorobromoethylene
CIP nomenclature: (E)-1-bromo-1,2,dichloroethene

Although the Cahn—Ingold—Prelog system has been extended to cover nearly all organic compounds, it is likely that the older D/L, cis/trans and other designations will continue to be used for some time.

It has already been pointed out that Fischer's assignment of the configuration of (+)-glyceraldehyde was purely arbitrary. The assignment had a 50:50 chance of being correct. It was not until 1951 that J. M. Bijvoet (appropriately from the Van't Hoff Laboratory at the University of Utrecht) was able to confirm experimentally the correctness of this assignment on (+)-sodium rubidium tartrate using anomalous X-ray diffraction techniques. (Ordinary X-ray diffraction cannot distinguish between enantiomers.)

10

The search for chiral centers other than carbon

As the Van't Hoff—Le Bel theory gained wider acceptance toward the end of the 19th century, chemists began to ask whether optical activity or stereoisomerism need be found only in carbon compounds or whether it could be associated with other elements as well. Initially this interest was confined to considerations of the stereochemistry associated with the presence of a nitrogen atom in either a trivalent state (amines, oximes) or 'pentavalent' state (ammonium salts). The investigations centered around the stereochemistry of compounds containing trivalent nitrogen will be discussed first.

A. TRIVALENT NITROGEN

In 1883, Heinrich Goldschmidt in Zürich reported the existence of two isomers of the dioxime of benzil ($C_6H_5COCOC_6H_5$). Little notice was taken of this until 1887 when E. O. Beckmann reported two isomeric oximes of an even simpler carbonyl compound, benzaldehyde. Beckmann suggested that the two compounds might be structural isomers having the following structures:

$$C_6H_5-CH{=}NOH \qquad C_6H_5-\underset{\underset{\displaystyle O}{\diagdown\diagup}}{CH-NH}$$

'Benzaldoxime'   'Isobenzaldoxime'

Goldschmidt, on the other hand, thought that they might be stereo-isomers. Victor Meyer and Karl Friedrich von Auwers, at the University of Göttingen, suggested that stereoisomerism for the benzil dioximes might arise because of a restricted rotation about the central carbon—carbon single bond. They also predicted the existence of a third isomer, which they synthesized shortly thereafter. The Meyer–Auwers theory, however, was unable to account for the existence of the isomeric benzaldoximes.

An alternative to these theories was proposed by Alfred Werner in 1890 in the theoretical portion of his doctoral dissertation at the Eidgenössisches Polytechnikum. Shortly thereafter, Wener's theory was published in the *Berichte der Deutschen Chemischen Gesellschaft*, with Arthur Hantzch as a co-author. Although the theory is usually referred to as the Hantzch–Werner hypothesis, Hantzch pointed out that the ideas concerning the stereochemistry of nitrogen were 'essentially the intellectual property of Herr Werner'. The stereoisomerism of the oximes was attributed by Werner to the geometry of the nitrogen valences:

> The three valences of the trivalent nitrogen atom (perhaps valences of other multivalent atoms also) do not always lie in a plane with the nitrogen atom itself. In certain compounds, the three valences of the nitrogen atom are directed toward the corners of a (in any case irregular) tetrahedron whose fourth corner is occupied by the nitrogen atom itself.

It might be noted that Werner placed the nitrogen at the *corner* of the tetrahedron rather than at the center. The Hantzch–Werner structure of hydrogen cyanide, $HC \equiv N$, is shown in Fig. 47a. The two-dimensional structures of the two benzaldoxime and the three benzildioxime stereoisomers used by Hantzch and Werner are shown in Fig. 48. The use of the *syn-, amphi-* and *anti-* prefixes was proposed by Hantzch in 1891.

For compounds (such as amines and hydrazine derivatives) in which there were no multiple bonds to nitrogen, Hantzch and Werner thought that the nitrogen valences were coplanar that '. . . could be deflected from this plane only by certain influences, e.g. by the mutual attraction or repulsion of the radicals bonded to the nitrogen atom'. A number of investigations were undertaken at the end of the century that were directed toward the isolation of stereoisomers of trivalent nitrogen compounds. In 1889, Le Bel attempted, without success, to isolate optically active methylethylpropylamine, $(CH_3)(C_2H_5)(C_3H_5)N$, from solutions of the amine that had been exposed to the actions of various moulds. It should be recalled (Chapter 5) that Le Bel's concept of the origin of molecular

Figure 47. Hantzch–Werner structures of (a) $HC \equiv N$, (b) and (c) *trans*- and *cis*-benzaldoximes.

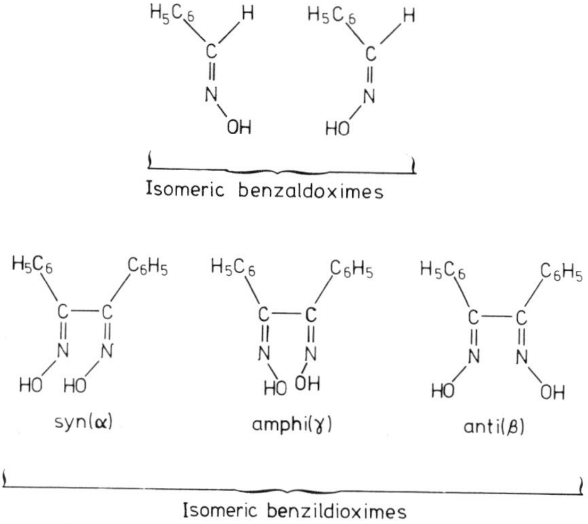

H_5C_6 H H_5C_6 H
    C   C
    ‖   ‖
    N   N
       OH   HO

Isomeric benzaldoximes

H_5C_6 C_6H_5 H_5C_6 C_6H_5 H_5C_6 C_6H_5
    C—C   C—C   C—C
    ‖ ‖   ‖ ‖   ‖ ‖
    N N   N N   N N
 HO HO   HO OH   HO   OH

   syn(α)   amphi(γ)   anti(β)

Isomeric benzildioximes

Figure 48. Hantzch–Werner formulas of stereoisomers of benzaldoxime and benzildioxime.

asymmetry was compatible with the undertaking of this investigation. Other investigators attempted resolutions of amines through the preparation of the tartrate salts. The failure of these resolutions was attributed by some to be related to the ease of hydrolysis of the salt. In an attempt to minimize this possibility, resolutions were attempted using salts prepared from strong acids, such as (+)-camphorsulfonic acid, and in a non-hydroxylic solvent such as ethyl·acetate. In 1904, F. S. Kipping and A. H. Salway at the University of Nottingham were unsuccessful in even separating the 'diastereomers' of a covalent derivative of a 'racemic' amine:

$$(+)\text{-}C_7 H_7 \overset{*}{C}H(CH_3)COCl + (``\pm")\text{-}C_6 H_5 NHCH_3 \longrightarrow$$

$$C_7 H_7 \overset{*}{C}H(CH_3)CON(CH_3)C_6 H_5$$

The persistently negative results obtained in these investigations only made it more difficult to understand why the non-coplanar geometry of the nitrogen valences in some compounds, such as the oximes, was not preserved in others. The Hantzch–Werner hypothesis received some additional experimental support in 1910 when W. H. Mills and A. M. Bain were able to resolve the oximes of cyclohexanone-4-carboxylic acid (Fig. 39, Chapter 8). Further reports of resolutions of a number of oximes and hydrazone compounds by Mills and his students in the next two decades amply confirmed the validity of the Hantzch–Werner hypothesis in compounds containing a carbon–nitrogen double bond.

The few examples of trivalent nitrogen compounds in which the three bonds could not be coplanar were found in unusual polycyclic compounds, such as quinuclidine, $N(CH_2 CH_2)_3 CH$, which contained nitrogen at a bridge head. By the 1930s, however, some physical property measurements of ammonia and amines suggested a non-planar geometry. For example, the measurable dipole movement of ammonia was not consistent with the three hydrogens residing in the same plane as nitrogen.

These studies, however, provided no clue as to why the amines could not be obtained in an optically active form. As early as 1924, Meisenheimer had suggested that a non-planar nitrogen compound, Nabc, might exist as a racemic mixture if it was imagined that the nitrogen could pass easily through the plane of the three attached groups (a, b, c) to produce its enantiomer. An interpretation of the absorption spectra of ammonia was found later to be consistent with this view, and a barrier to inversion was calculated to be in the range of $6-11$ kcal/mole. Thus the resolution of an amine would only be successful at room temperature if the barrier were greater than about 20 kcal. A low temperature ($-78\ ^\circ C$) resolution of N-ethylaniline was attempted, without success, in 1932. Resolutions of cyclic amines were attempted by a number of investigators following calculations that estimated that the inversion barrier in such compounds might be as high as 38 kcal/mole. In more rigid bicyclic heterocyclic systems, such as is illustrated by 'Troger's base' (Fig. 49), the nitrogen inversion is not possible and the enantiomeric form should be stable. The resolution of Troger's base was first reported by V. Prelog and W. Wieland in 1944.

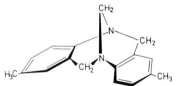

Figure 49. 'Troger's base' (2,8-dimethyl-6H,12H-5,11-methanodibenzo [b, f] [1,5] diazocine).

It was not until the 1950s that nuclear magnetic resonance measurements of monocyclic amines (by A. T. Bottini and J. D. Roberts in 1956) not only provided evidence of the non-planarity of the three groups bound to nitrogen, but also indicated that the rate of inversion was so large that it would not be possible to resolve such compounds even at low temperatures. In 1968, S. J. Brois showed by NMR studies that the rate of inversion in the C-substituted N-haloaziridines might be low enough to allow the separation of *cis/trans* isomers. Shortly thereafter the seaparation of two isomers (Fig. 50) by means of gas chromatography was reported.

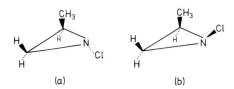

Figure 50. Diastereomeric N-chloroaziridines.

B. 'PENTAVALENT' NITROGEN

The reader might find it difficult to understand how it was that the stereochemistry of trivalent nitrogen was not related more closely to the stereochemistry of ammonium salts. It should be mentioned, first of all, that prior to about the 1870s, ammonium salts (such as $NH_4 Cl$) were considered as 'molecular complexes'. When the salt was considered as a molecular complex, the formula might be written $NH_3 \cdot HCl$. (This way of talking about salts still survives today in non-systematic chemical nomenclature; for example, 'ephedrine hydrochloride'.) This manner of formula writing derived from the views of Kekulé who maintained that the valency of every element should be constant. In ammonium chloride, for example, nitrogen was still considered to be trivalent, the hydrogen chloride being bound to the ammonia by a 'secondary' attractive force. It was difficult to reconcile this theory with August Hofmann's demonstration (in 1851) of the equivalency of the four alkyl groups in quaternary ammonium salts by the following reaction scheme:

$$(C_2 H_5)_3 N + C_5 H_{11} I \longrightarrow (C_2 H_5)_3 (C_5 H_{11}) NI \xrightarrow[OH^-]{heat}$$

$$(C_2 H_5)_2 (C_5 H_{11}) N + C_2 H_4$$

It would not be possible to account for the formation of ethene ($C_2 H_4$) unless the four alkyl groups were equivalent since the heating of pentyl iodide ($C_5 H_{11} I$) with sodium hydroxide would produce pentene. This result suggested that the valency of nitrogen might indeed be variable – in this case having a valency of four or five. The equivalency of the alkyl groups in quaternary ammonium salts was again demonstrated experimentally in 1875 by Victor Meyer and M. Lecco. They noted, for example, that the decomposition of a quaternary salt, prepared by two synthetic routes, produced the same products:

$$(CH_3)_3 N + C_2 H_5 I \longrightarrow (CH_3)_3 (C_2 H_5)NI \longleftarrow (CH_3)_2 (C_2 H_5)N + CH_3 I$$

$$\text{heat} \Big| OH^-$$

$$(CH_3)_3 N + C_2 H_4$$

The advent of stereochemistry stimulated several suggestions concerning the three-dimensional structures of 'quinquevalent' (pentavalent) nitrogen compounds (Fig. 51). In the first two structures, groups attached at positions 1 through 3 are equivalent and were originally derived from the parent amine. Positions 4 and 5 are occupied by hydrogen and the 'acid radical' (halide) in non-quaternary salts, such as $(CH_3)_3 NHCl$. It was noted, however, that positions 4 and 5 were not equivalent in the cubical structure. In the tetragonal (or 'square') pyramid, position 5 is occupied by the halide group. All of these structures soon came under critical scrutiny since more stereoisomers were predicted than was experimentally confirmed (Table 1).

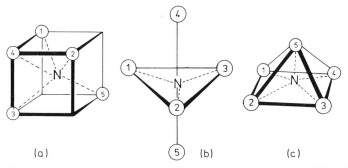

Figure 51. Structures of pentavalent nitrogen proposed in 1878–90: (a) cubical (Van't Hoff, 1878), (b) trigonal bipyramid (Willgerodt, 1888; Burch and March, 1889), (c) tetragonal pyramid (Bischoff, 1890).

In contrast to Van't Hoff, who had suggested in 1878 that ammonium salts had cubical structures, Le Bel thought that any particular molecular configuration would have only a transitory existence due to the repulsive forces between the groups attached to the nitrogen, although the presence of more complex groups might increase the stability to allow the isolation of the substance in a particular geometry. In pursuit of this idea, Le Bel in 1891 attempted to obtain optically active samples of a number of ammonium salts. The fact that he had earlier observed two crystalline forms of trimethylisobutylammonium platinochloride probably suggested to him that even ammonium compounds having identical alkyl groups might be asymmetric. Although none of the above compounds could be

TABLE I

	Predicted number of stereoisomers		
Type of compound	Cubical	Trigonal bipyramid	Tetragonal pyramid[a]
A_3BNX	2	2	1
A_2B_2NX	2	2	2
A_2BCNX	4[b]	4[b]	2
ABCDNX	8[c]	8[c]	6[c]

[a] Halide (X) assumed to be in position 5.

[b] One enantiomeric pair.

[c] All stereoisomers as enantiomeric pairs.

obtained in an optically active form, Le Bel did report that a small optical rotation was produced in the solutions containing isobutylpropylethyl-methylammonium chloride which had been exposed to the action of the *Penicillum glaucum* mould. When Willy Marckwald reported that he was unable to confirm this result in 1899, Le Bel repeated his experiment and maintained the correctness of the result. It was only in 1912 that Pope and Read provided evidence that the optical activity reported by Le Bel was probably due to an impurity in the solution. By this time, however, other investigators had demonstrated the existence of asymmetric quaternary ammonium salts.

A number of chemical resolutions of quaternary ammonium compounds were attempted in the 1890s without success. The English chemists, William Jackson Pope and S. J. Peachey, reported the successful resolution of the *d*-camphorsulfonic acid salt of allylbenzylmethylphenyl-ammonium iodide from the non-hydroxylic solvents acetone and ethyl-acetate. A pentavalent three-dimensional structure was suggested, although no illustrations were included:

> . . . it is proved that quaternary ammonium derivatives in which the five substituting groups are different, contain an asymmetric nitrogen atom, which gives rise to antipodal relationships

of the same kind as those correlated with an asymmetric carbon atom.

Although the salt-like character of the ammonium compounds was more generally recognized by the end of the century, many chemists continued to discuss the stereochemistry of the ammonium compounds in terms of a pentavalent configuration. The Bischoff structure (Fig. 51c) seemed to enjoy the greatest popularity. In 1903, H. O. Jones concluded that none of the space-filling models of ammonium compounds were satisfactory, since he had been unsuccessful in his attempts to resolve salts of the formula, $R^1 R^2 R_2^3 NX$.

It is to Alfred Werner again that we must credit the origins of our present view of the structure and stereochemistry of ammonium compounds, although he did not consider the geometrical orientation of the nitrogen valences in his 1890 paper. In a paper published in 1893, Werner not only argued that the nitrogen should be tetravalent but should have a tetrahedral geometry:

> In accordance with this behavior of carbon, it may be assumed that for boron and nitrogen, which likewise have coordination number four, the four coordination positions will also be found in the relative arrangement of the corners of a tetrahedron.

The halogen was bound to the nitrogen by an ionizable bond which did not affect the stereochemistry of the ammonium ion. Werner's ideas on the constitution of the ammonium salts were not well known or appreciated until he republicized them in papers and books published in the first two decades of the 20th century. The difficulty that some chemists had in understanding Werner's earlier views is illustrated in the comments of S. V. Pickering in 1893:

> The absurdity of the tetrahedron conception becomes more glaring when we pass from triad to pentad nitrogen compounds. The nitrogen atom is promptly taken away from the corner, and replaced in the middle of the tetrahedron; but as this still leaves one hydrogen too many, this hydrogen is crammed into the middle to keep the nitrogen company, $(x-N)abcd$ being considered as a perfect analogue of Cabcd. Such playing fast and loose with the atoms, and making them into a four-cornered figure at any cost, is scarcely a scientific mode of dealing with the question.

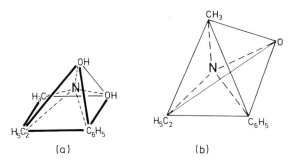

Figure 52. Possible configurations of ethylmethylphenyl N-oxide, EtMePhNO: (a) tetragonal pyramidal (for monohydrate), (b) tetrahedral (anhydrous compound).

Experimental support for the tetrahedral hypothesis came from the investigations of a different class of nitrogen compounds. In 1908, Jakob Meisenheimer in Berlin prepared optically active methylethylphenyl N-oxide (MeEtPhN=O). When the N-oxide was treated with hydrochloric acid, an optically active addition product [MeEtPhN(OH)Cl] was formed. With respect to this compound, Meisenheimer concluded that 'the four valences of a quinquevalent nitrogen were directed towards the angles of a tetrahedron, the fifth (ionizable) valency being very mobile'. The tetrahedral orientation of the four groups attached to the N-oxide could not be definitely established at that time since the compound produced on the addition of barium hydroxide to the acidic solution could not be obtained in a crystalline or anhydrous state. It was considered possible, therefore, that the N-oxide existed as the hydrate, in which the five bonds formed a tetragonal pyramid (Fig. 52a). By 1911, however, Meisenheimer had prepared the crystalline anhydrous compound which supported a tetrahedral configuration (Fig. 52b). Meisenheimer's and Werner's views were receiving a sympathetic treatment as can be seen from the discussion found in the *Annual Reports of the Chemical Society*:

> The four non-ionizable radicals group themselves about the central nitrogen atom in a first sphere of attraction in the relative positions of the angles of a tetrahedron . . . whilst the ionizable mobile group is more loosely attached in a second outer zone in some position opposite to the center of one of the tetrahedron faces. For N-oxides, the oxygen takes the place of one of the four non-ionzable radicals and the anion.

(It is not clear how the oxygen could occupy a position on one corner of the tetrahedron *and* the place occupied by the anion. The geometry of the ammonium ion suggested by the above passage is illustrated in Fig. 53.)

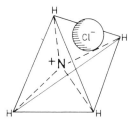

Figure 53. Stereochemical formula of NH_4Cl (as probably visualized in about 1910–20).

That the anionic group in the quaternary ammonium salts was not exchangeable with the alkyl groups was demonstrated experimentally by a number of investigators. For example, Schlenk and Holtz reported in 1916–17 that no exchange among the alkyl and aryl groups could be observed in solutions of $(CH_3)_4N^+ (C_6H_5)_3C^-$ or $(CH_3)_4N^+ C_6H_5CH_2^-$. A decade later, Hager and Marvel found that the replacement of the halide ion by an alkyl group did not produce an equivalency among the five alkyl groups. The alkene and amine produced on the decomposition of the resulting quaternary salt was never derived from the 'anionic' alkyl group introduced in the first reaction:

(1) $R'Li + R_4N^+Br^- \rightarrow R_4NR'$

(2) $R_4NR' \xrightarrow[\text{heat}]{OH^-}$ alkene (derived from R) + R_3N

The Bischoff structure was finally shown to be untenable in 1925 when W. H. Mills and E. H. Warren obtained 4-phenyl-4'-carbethoxy-bis-piperidinium-1,1-spirane bromide (Fig. 54) in an optically active form. If the valences were in a pyramidal form, the structure would be achiral (since a plane of symmetry can be located that bisects the nitrogen atom and the 4,4'-carbons).

By the late 1920s, the formulas of ammonium compounds were written in terms of the electronic theory of valency proposed by G. N. Lewis in 1916. The 'pairing' of valence electrons to form a covalent bond was first represented by Lewis in 1916 in terms of a 'cubical' model. A single bond

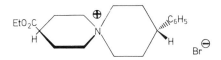

Figure 54. 4-Phenyl-4'-carbethoxy-bis-piperdinium-1,1-spirane bromide.

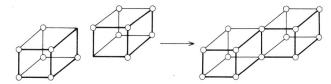

Figure 55. Use of G. N. Lewis's 'cubical' models to illustrate the formation of a single bond.

involved the sharing of a pair of valence electrons on the edges of two cubes (Fig. 55). A double bond would correspondingly involve the sharing of two pairs of electrons at the face of two cubes. The limitations of the cubical model were apparent to Lewis when he attempted to discuss the carbon–carbon triple bond. It was also difficult to reconcile the tetrahedral geometry of the ammonium ion with the cubical disposition of valence electrons:

> In order to illustrate this point [the sharing of a pair of electrons in a single bond] we may discuss a problem which has proved extremely embarrassing to a number of theories of valence. I refer to the structure of ammonia and of ammonium ion. The ammonium ion may, of course, on account of the extremely polar character of ammonia and hydrogen ion, be regarded as a loose complex due to the electrical attraction of the two polar molecules. However, as we consider the effect of substituting hydrogen by organic groups we pass gradually into a field where we may be perfectly certain that four groups are attached directly to the nitrogen atom, and these groups are held with sufficient firmness so that numerous stereochemical isomers have been obtained. . . . The chloride in ammonium
>
> chloride (H:N̈:H$^+$:|C̈l:$^-$)   is not attached directly to the
>
> nitrogen but is held simply through electronic forces by the ammonium ion.
>
> Assuming now, at least in such very small atoms as that of carbon, that each pair of electrons [involved in the four single bonds] has a tendency to be drawn together, perhaps by magnetic force if the magneton theory is correct, or perhaps by other forces which become appreciable at small distances, to occupy positions indicated by the dotted circles (Fig. 56). We then have a model which is admirably suited to portray all of the characteristics of the carbon atom.

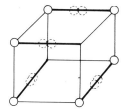

Figure 56. G. N. Lewis's tetrahedral 'pairing' of electrons.

The carbon—carbon triple bond could be produced by the sharing of three pairs of electrons at the face of two tetrahedra. Although Lewis did consider electronic structures of ammonia and amines, he did not discuss their geometrical structures. A few years after Lewis published his ideas about the structure of the ammonium ion, Nevil Vincent Sidgwick used the ammonium ion as the simplest example of a 'coordinate' bond, a bond in which one group of atoms ($H_3 N:$) provides both of the bonding pairs of electrons to another ion (H^+) or group.

Lewis's 'cubical' model enjoyed only a short popularity, most notably in the writings of Irving Langmuir, since the model was not compatible with the stereochemical information. One critic could only say that the '. . . atoms lay around like dry-goods boxes in a packing shed'.

C.   OTHER HETEROATOMS

Until the early part of the 20th century, the sulfur in sulfonium compounds ($R_3 SX$) was regarded as a 'quadrivalent' element. It should come as no surprise to learn that a tetrahedral orientation of the four groups about sulfur was suggested after the isolation of optically active forms of ethylmethylphenacyl bromide (Fig. 57a) by Samuel Smiles and thetaine chloride (Fig. 57b) by W. J. Pope and S. J. Peachey in 1900. The latter investigators also were able to obtain an optically active tin compound in the same year. (In 1900, ionic structures like those shown in Fig. 57 were not used; instead four bonds connected the central sulfur atom to the four substituents.) These reports stimulated a number of investigations

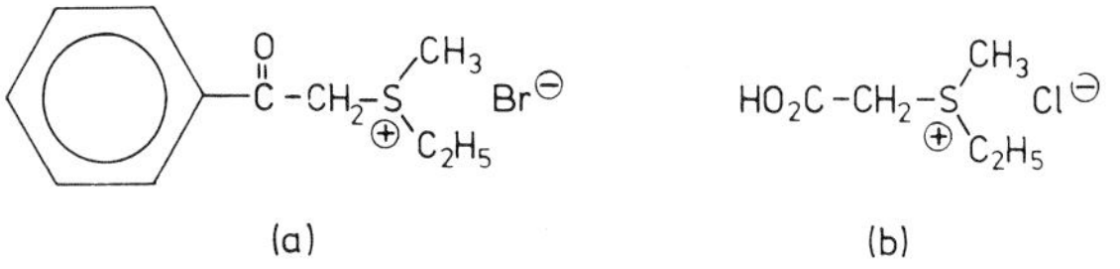

Figure 57. Chiral sulfonium compounds prepared in 1900: (a) ethylmethylphenacyl bromide (S. Smiles), (b) thetaine chloride (W.J. Pope and S.J. Pechey).

during the next few decades which were directed toward the isolation of stereoisomers that owed their asymmetry to the presence of some hetero-atom.

The introduction of electronic theories of bonding in the 1920s re-quired a re-evaluation of the cause of the particular geometry proposed for the 'quadrivalent' sulfur compounds. It was not yet clear in the early 1920s whether the sulfonium compounds (Fig. 57) should be considered 'quadrivalent' or 'trivalent'. Considerable interest was generated, there-fore, by the publications of Henry Phillips in 1925–26 who reported the resolution into enantiomers of a sulfinic ester and two sulfoxides (Fig. 58a, b). These were the first obvious examples of asymmetric compounds containing a central heteroatom bonded to only three groups. Since no evidence was available of optically active ketones, Phillips ruled out the existence of the sulfur–oxygen double bond as written in the structures 57a, b. Instead the bond must be a 'semipolar' one (that is, a coordinate covalent bond). In the tetrahedral structures provided by Phillips (Fig. 57c, d), the sulfur atom was placed at the apex of a tetrahedron. The non-bonding electrons (indicated by 'XX') thus did not formally occupy the corner of a tetrahedron, as it is presently considered. (It might be noted that in 1927, Sidgwick provided illustrations of asymmetric sulfur com-pounds in which he placed the sulfur in the center of a tetrahedron – one corner of the tetrahedron was left unoccupied by either a group or electron pair!)

The resolution of a number of octahedral complexes of inorganic com-pounds by Alfred Werner in 1911 provided dramatic evidence of the importance of considering the symmetry of the molecules as a whole. Any lingering doubts that the optical activity of these complexes might some-how be due to the presence of organic groups was put to rest by Werner's resolution of a completely inorganic complex of cobalt in 1914. (See Kauffman, *Inorganic Chemical Complexes*, in this series.)

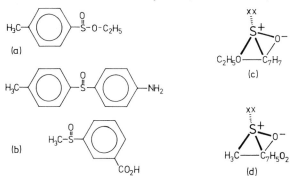

Figure 58. Chiral tetravalent sulfur compounds prepared by H. Phillips, 1925–26.

11

The stereochemistry of trivalent carbon species

Up to now we have considered the stereochemistry of substances that are isolatable in the laboratory. When chemists considered the geometries of species that might be presumed to exist for only a moment in a chemical transformation, the study of stereochemistry had begun to move from what might be characterized as a 'static' phase to a 'dynamic' phase. New problems are confronted and investigative techniques are called for when the chemist asks how you can learn about the stereochemistry of a system that will not allow him to observe it. This chapter summarizes some of the problems concerned with proposals related to the geometry of trivalent carbon species.

In terms of present bonding theory, the four covalent bonds to carbon are considered to involve four bonding orbitals containing two electrons each. When one of these orbitals is not engaged in a bond, the carbon atom is considered tricovalent or trivalent. Depending on whether the non-bonding orbital is empty, or contains one or two electrons, we are concerned with the carbonium ion, or carbocation, $(R_1 R_2 R_3 C \oplus)$, a free radical $(R_1 R_2 R_3 C \cdot)$, or a carbanion $(R_1 R_2 R_3 C: \ominus)$. In the older stereo-chemical literature, all of these species were considered to be derived from a tetrahedral carbon atom by leaving one apex of the tetrahedron non-bonded. Whether the tetrahedral geometry was preserved when these trivalent species were formed was not a question that was easily answered, and the difficulties encountered were not unlike those associated with the stereochemistry of trivalent nitrogen.

The interest in the stereochemistry of these species first had to await the appearance of evidence that indicated their existence. In 1900 Moses Gomberg, at the University of Michigan, furnished the first experimental evidence of a stable free radical. (It should be pointed out, however, that the formulas written by Gomberg did not show any unpaired electron on the trivalent carbon atom.)

A. CARBONIUM IONS (CARBOCATIONS)

The salt-like character of the triphenylmethyl halides (Ph_3CX) used by Gomberg in his preparation of the free radicals was soon recognized, but further experimental studies were required before a trivalent cationic species was considered either as a stable entity or an intermediate in chemical reactions. The investigations related to the Walden inversion in which a carbonium ion intermediate was proposed have already been mentioned (Chapter 7) and it is difficult to separate these studies from the present discussion.

In 1912, Biilman suggested that the formation of optically active lactic acid in the reaction of silver oxide with optically active 2-bromo-propanoic acid might involve the intermediate formation of a dipolar intermediate, $CH_3 - \overset{+}{C}H - CO_2^-$. The formation of the optically active product on the reaction of this intermediate with water could only be explained if the positively charged carbon retained the asymmetry found in the 2-bromopropanoic acid. Although most chemists could see no reason why the carbon should retain its asymmetry, no persuasive evidence of the stereochemistry of such a species could be obtained until it could be demonstrated that such an intermediate was involved in the reaction.

The kinetic and mechanistic studies of Hughes and Ingold (Chapter 7) provided a basis, however, on which to decide whether a substitution reaction might take place by a simultaneous process, or a stepwise one involving an intermediate carbonium ion. The fact that racemization invariably accompanied substitution reactions that proceeded by the second mechanism strongly argued for the intermediacy of a planar carbonium ion. Since the racemization was not always complete, Hughes and Ingold commented:

> Unimolecular substitution in alkyl compounds leads normally to racemization, because the intermediate ion has a plane of symmetry; but if the life of the ion is short racemization will be incomplete and will accompany a predominating inversion, owing to the circumstances that the separating ions shield each other . . .

The possibility that the ion might be able to retain its asymmetry in some cases was not ruled out, however:

> A charged substituent of neutralizing sign (e.g. a carboxylate-ion substituent in a cation) may stabilize a pyramidal configuration in the ion, and lead to eventual substitution with retention of form:

$$\begin{array}{c} CO \\ \diagup \quad \diagdown \\ {}^{-}O \qquad C^{+}\cdots Br^{-} \end{array}$$

By the 1930s a number of papers had been published designed to provide experimental tests of the geometry of the carbonium ion. Although Gomberg, Wallis and Adams had reported that a racemic product was invariably produced in reactions which involved an intermediate carbonium ion, it still was not possible to decide whether the carbonium ion was planar or rapidly equilibrating between enantiomeric tetrahedral configurations. In 1939, Paul Bartlett and L. H. Knox, at Harvard University, found that apocamphyl chloride (Fig. 35a, Chapter 7) was inert to $AgNO_3$ solution, under conditions in which the non-bridged tertiary alkyl chloride (35b) reacted instantaneously. The stability of these and related compounds indicated the difficulty of forming a non-planar carbonium ion. However, later studies (Chapter 16) have shown that the reluctance of these compounds to form carbonium ions could be attributed in part to the fact that the ion requires a back-side solvation.

At about the time when the requirement for the coplanarity of these 'normal', or 'classical', carbonium ions was generally accepted, we find the first proposals for the existence of a species of carbonium ion of decidedly unusual geometry. These were subsequently termed 'non-classical' carbonium ions. In the early literature of the 1940s, the investigations of these ions were associated with the names of several organic chemists in California: J. D. Roberts, Saul Winstein and Donald J. Cram. Thomas P. Nevell, Eduardo de Sales and Christopher L. Wilson at University College, London, are credited with the first proposal of such an ion in 1939 in order to account for the stereochemistry of the Wagner–Meerwein rearrangement. It was proposed that the ionization of camphene hydrochloride (a) might involve the formation of an intermediate ion (b) that was 'mesomeric' between the camphene (a) and isobornyl (c) structure, rather than a rapidly equilibrating mixture of the 'classical' ions (d) and (e):

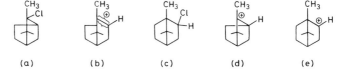

(a)   (b)   (c)   (d)   (e)

In several examples of 'non-classical' carbonium ions illustrated in Figs. 59 and 60, the positive charge is found to be delocalized over several carbon atoms, some of which are not covalently bound to each other. The delocalization was thought to be produced by the overlap of the vacant orbital of a carbonium ion (Fig. 59a, c) with some non-conjugated and

(a)   (b)   (c)   (d)

Figure 59. Representations of the 'classical' (a, c) and 'non-classical' (b, d) carbonium ions: the norbornyl ion (a, b); the norbornenyl ion (c, d).

distant σ- or π-bond orbitals within the same molecule to produce a special stabilization of the ion (Fig. 59b, d). The bicyclobutonium ion (Fig. 60) was postulated as an intermediate by J. D. Roberts and R. H. Mazur in 1959. This non-classical ion might be considered the resonance hybrid of the three carbonium ions shown in the figure. The structures of the three ions do not adequately illustrate the spatial arrangement of the carbon skeleton.

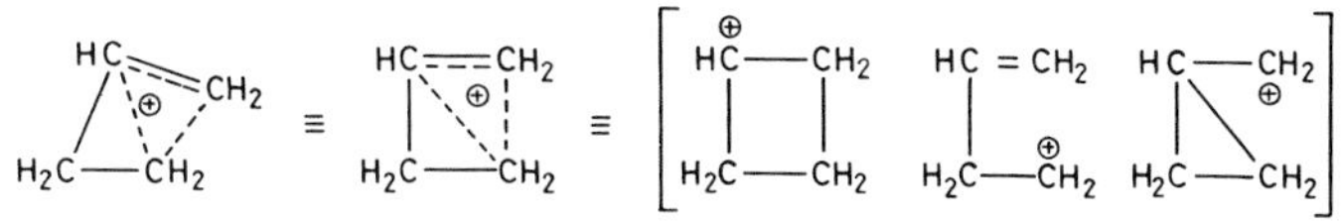

Figure 60. Representations of the bicyclobutonium ion.

The intermediacy of such ions was usually proposed to account for the observation of unusual rate effects or stereochemistry of solvolysis reactions. The σ- or π-bonds were, therefore, thought to exert an 'anchimeric (or synartetic) assistance' in bond breakage. The proposal of ions of such unusual geometry is perhaps not so intuitively unreasonable, if the stereochemical consequences of other reactions (pinacol rearrangement, the Walden inversion) are considered in terms of transition states involving a 'pentavalent' carbon.

The enthusiasm for proposals of such intermediates reached a high point in the 1960s, often to the point where non-classical ions were proposed even though the experimental evidence did not require it. A 1964 issue of *Chemical Abstracts* contains an abstract that illustrates the extremes to which at least some chemists might go:

> Only a dark, undistillable resin remained upon removal of the solvent. This suggests that a nonclassical carbonium ion intermediate is involved in the mechanism.

B.   CARBON 'FREE RADICALS'

It was not until the 1930s that experimental work was undertaken to determine the geometry of the carbon 'free radicals'. Roger Adams and

E. S. Wallis concluded that the free radical could not maintain the asymmetry present in the compound from which it was derived. They based their conclusion on the observation of the formation of an optically inactive solution when $(-)$-phenyl-p-biphenyl-α-naphthylthioglycollic acid was heated with the triphenylmethyl radical.

The existence of an asymmetric radical is suggested by the report of Karagunis and Drikos in 1933 who found that the reaction of chlorine with phenylbiphenyl-α-naphthylmethyl in carbon tetrachloride solution in the presence of circularly polarized light produced an optically active product (triaryl chloride). In 1949, Karagunis attempted, without success, to obtain an optically active triarylmethyl radical by chromatography on several optically active absorbants.

Although most spectroscopic studies were more compatible with the radicals having a planar configuration, it was also clear that a non-planar configuration could be more easily attained than with the carbonium ion. For example, a reaction studied by Paul Bartlett in 1954 involved the relatively facile formation of a non-planar, bridged free radical:

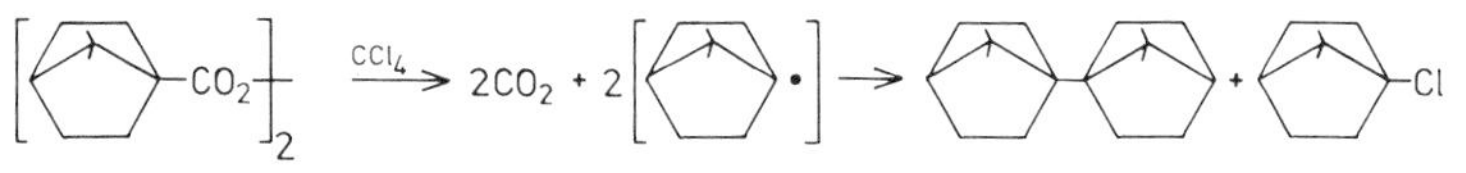

C.   CARBANIONS

The stereochemistry of the carbanion might be considered to be more analogous to the stereochemistry of amines, since the carbon and nitrogen in both are isoelectronic. The normal means of producing a carbanion involves the removal of a proton from a tetravalent carbon by a strong base. Since such protons are only removed with difficulty, the compounds studied are those that contain substituents that stabilize the resulting carbanion. Stabilization of the carbanion by resonance would favor a planar configuration. Some of the earliest stereochemical studies were of the carbanions derived from nitroalkanes. Before the advent of resonance theory, it was thought that the anionic forms (a) and (b) in Fig. 61 should be capable of independent existence. In 1927, R. Kuhn and H. Albrecht reported that an optically active sodium salt could be prepared from optically active 2-nitrobutane and that the nitroalkane formed on reacidification still exhibited a slight optical activity. These results were interpreted in terms of the formation of the carbanion species (a) and not the *aci*-anion (b). It was not until 1947 that Nathan Kornblum found that the presence of an optically active impurity, $CH_3 CH_2 CH(CH_3)ONO_2$, was responsible for the optical activity of the salt observed by Kuhn. Ambiguous results were obtained in other studies undertaken in the 1930s

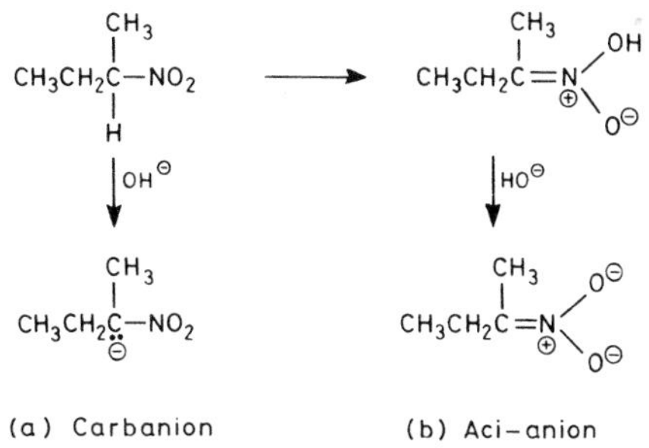

$$\underset{H}{\overset{CH_3}{CH_3CH_2\overset{|}{\underset{|}{C}}-NO_2}} \longrightarrow CH_3CH_2C\overset{CH_3}{=}\overset{OH}{N_{\oplus}}\overset{OH}{\diagdown O^{\ominus}}$$

$$\Big\downarrow OH^{\ominus} \qquad\qquad\qquad \Big\downarrow HO^{\ominus}$$

$$\underset{\ominus}{\overset{CH_3}{CH_3CH_2\overset{|}{\underset{|}{C}}-NO_2}} \qquad\qquad CH_3CH_2C\overset{CH_3}{=}\overset{O^{\ominus}}{N_{\oplus}}\overset{O^{\ominus}}{\diagdown O^{\ominus}}$$

(a) Carbanion   (b) Aci-anion

Figure 61.

to ascertain the existence of an asymmetric carbanion. More systematic studies were not undertaken until the 1960s. These studies revealed that the optical purity of the products formed from the carbanion depended on a number of factors — such as the solvent used or the nature of the metal ion associated with the carbanion. Donald J. Cram, who was one of the most active investigators in this field, found that the hydrogen–deuterium exchange of optically active α-phenylethyl-α-d-methyl ether with potassium t-butanol proceeded with 94% retention of configuration with each act of exchange:

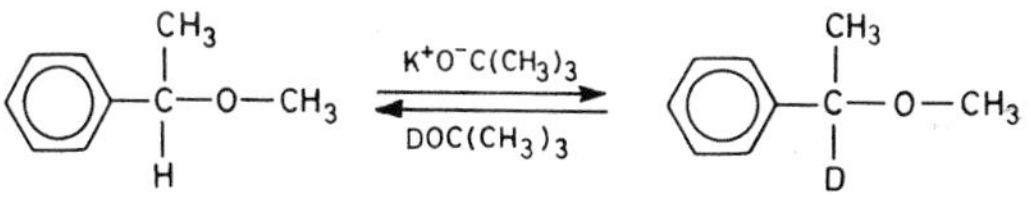

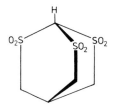

When, however, a more highly ionizing solvent, such as dimethylsulfoxide, was used, the exchange was accompanied by considerable racemization. That the carbanion need not exist in a planar configuration can be seen from the report of William von E. Doering in 1955 who found that the bridgehead hydrogen in a bicyclic sulfone (Fig. 62) could be removed by even the weak base, sodium bicarbonate.

Figure 62.

12

Rotational 'barriers' about single bonds and steric effects

A. ROTATIONAL BARRIERS

It is sometimes supposed that organic chemists of the late 19th and early 20th century generally assumed that there was 'free' rotation about the C–C single bond, a view that was only seriously challenged when the results of a number of physical and spectroscopic measurements were published in the 1930s and 1940s. The concept of the 'free' rotation is also thought to have been proposed by Van't Hoff to account for the non-existence of an excessive number of (conformational) isomers. That this simplifies Van't Hoff's own position has already been commented upon in Chapter 5.

In point of fact, a number of chemists in the latter part of the 19th century did attempt explanations of the stereochemistry of some reactions in terms of 'favored' conformations of the molecules. For example, Victor Meyer and E. Riecke in 1888 attempted to account for the existence of benzil dioxime isomers (see Chapter 10) by proposing that the rotation about the carbon–carbon single bond was restricted by the electrical valency 'dipoles' of the various groups about the carbon atoms. The publication of the Hantzch–Werner hypothesis in 1891, however, discouraged further speculations about rotational barriers.

A more persuasive experimental basis in support of preferred configurations of molecules was provided in the 1890s by C. A. Bischoff at Riga. In what was described as his 'dynamical hypothesis', Bischoff argued that the tetrahedral angles of the carbon atoms in the straight-chain alkanes would favor a conformation which would place the fifth and sixth carbon atoms in close proximity to the first carbon atom. Substituents on these carbons should, therefore, affect the stability of this conformation. In succinic acid, for example, the most favorable conformation would be that which minimized the repulsive forces between the two carboxyl

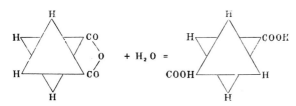

Figure 63. Hydrolysis of succinic anhydride to form succinic acid (in its
most 'favorable' conformation).

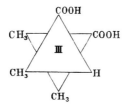

Figure 64. Bischoff's illustration of the 'favored' conformation of tri-
methylsuccinic acid, $HOOCCH(CH_3)C(CH_3)_2COOH$.

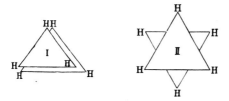

Figure 65.   Bischoff's formulas of the two conformations of ethane.

groups (Fig. 63). To illustrate the 'favorable' or 'unfavorable' conformations, Bischoff used a type of projection formula similar to the Newman
projection formulas introduced in the 1950s. (P. H. Hermans also employed similar formulas in the 1920s.) The observation that alkyl-substituted
succinic acids formed cyclic anhydrides more easily than succinic acid
was attributed to the alkyl–carboxyl repulsions being more important
than the carboxyl–carboxyl repulsions. Thus the most favorable conformation of 2,2,3-trimethylsuccinic acid placed the carboxyl groups
in closer proximity than in succinic acid (Fig. 64). It is of interest to
note that Bischoff thought that even in a simple compound such as
ethane, the repulsive interactions between the hydrogens should favor
the staggered over the eclipsed conformation (Fig. 65). Bischoff's
report of the isolation of two isomers of diethylmethylsuccinic acid,
$HO_2CC(C_2H_5)_2CH(CH_3)CO_2H$, originally provided the most persuasive
experimental evidence of the possibility of the independent existence of

rotational isomers. Unfortunately, perhaps, it was later shown that the two isomers were structural isomers.

By the turn of the century, most organic chemists felt that the experimental evidence then available indicated that the rotation about single bonds was for all practical purposes 'free', and that further speculations about the repulsive or attractive forces between the non-bonding groups that might stabilize particular rotational configurations could not be subjected to any experimental verification. The subtle experimental anomalies (the ease of cyclic anhydride formation or differences in ionization constants, for example) that had been reported, therefore, remained unexplained for another 50 years. In the 1920s, experimental evidence had been obtained from another area of research that reopened the question. This research did not, however, evolve as a result of the continuation of the earlier studies, although ultimately, as discussed in Chapter 14, the argument came back to its original form.

B.   STERIC HINDRANCE

In 1872, August Hofmann reported that a number of aniline derivatives containing methyl substituents in the 2,6-positions gave low yields of the quaternary ammonium compounds when they were heated with methyl iodide. He concluded, rather vaguely, that 'this inability to unite with methyl iodide must depend upon some kind of molecular arrangement'. The reason for the inertness of the compounds remained unexplained for some time, although a similar effect was noted in the 1880s by F. Kehrmann who found that oxime formation was prohibited in *ortho*-substituted quinones. Kehrmann thought that the *ortho* substituents must 'fill up the space' near the reaction center.

In the 1890s, Victor Meyer and his students undertook detailed studies to ascertain the relationship between the position of a substituent on an aromatic ring and its effect upon a reactive center. The studies were prompted by the observation that mesitylene carboxylic acid (2,4,6-trimethylbenzoic acid) could not be esterified by heating in a methanol—hydrochloric acid solution, conditions under which most other derivatives of benzoic acid produced high yields of the methyl ester. Based on the yields of the methyl esters isolated for a series of substituted benzoic acids (heated for 2 hours in a 2% $HCl-CH_3OH$ solution), Meyer concluded:

> When the hydrogen atoms in the two *ortho* positions to the carboxyl in a substituted benzoic acid are replaced by radicals such as Cl, Br, NO_2, CH_3, COOH, an acid results which can only be esterified with difficulty or not at all.

Subsequently, this statement was referred to as the 'law of esterification'. The inertness of *ortho*-substituted acids toward esterification was attributed to the difficulty of forming a tetravalent intermediate (A) due to the close proximity of the *ortho* substituent:

$$ArCOOH + CH_3OH \longrightarrow [ArC(OH)_2OCH_3] \longrightarrow ArCOOCH_3 + H_2O$$
$$(A)$$

To account for the formation of the methyl ester of di-*ortho*-substituted benzoic acids when the silver salt of the acid was treated with methyl iodide, it was argued that the silver could force the two *ortho* groups far enough apart to make the carboxyl group accessible for reaction. In 1897, Alexander Kellas, working in Meyer's laboratory, used kinetic studies to demonstrate that even a single *ortho* substituent could retard the rate of esterification. At the University of Nottingham, John Sudborough (who had done much of the earlier work with Meyer) undertook an investigation of the esterification of substituted aliphatic carboxylic acids with a view to ascertain whether the 'esterification law' applied only to aromatic compounds. Since the steric effects he observed were not very pronounced, he could only conclude:

> At present, we do not understand why the two atoms in the *ortho*-positions have an influence so much greater than that of three atoms attached to the carbon atom to which the carboxylic group is directly united; the explanation may perhaps be found in the stereochemistry of the benzene molecule.

For most chemists, Meyer's esterification law simply meant that it was not possible to prepare esters of di-*ortho*-substituted benzoic acid. In 1908, however, M. A. Rosanoff and W. L. Prager found that such esters could be prepared, without an acid catalyst, when the esterification was run at higher temperatures (sealed tube, about 183 °C, for 100 hours). It was also noted that for the di-*ortho*-substituted benzoic acids, the equilibrium constant for the esterification, far from being negligible, was of the same magnitude as benzoic acid itself. Thus the earlier failure to isolate the esters of di-*ortho*-substituted acids could be attributed to a rate difference and not using a sufficiently large excess of alcohol. The importance of the nature of the substituent on the yield could also be demonstrated: for example, 2,4,6-tribromobenzoic acid was esterified eight times more rapidly than 2,4,6-trinitrobenzoic acid. Rosanoff and Prager, however, were unable to account for the rate difference, suggesting only that the substituent effect should not be attributed to a 'mechanical' cause.

By the 1930s it had become more fashionable to explain the variations in rate effects as arising from the electronic effects exerted by the substituents, and accordingly there was a reluctance to consider that the 'bulk' of the group might also be important. For instance, H. B. Watson in the 1940s suggested that a 'chelation' of the *ortho* methyl group and the carbonyl oxygen in the transition state might affect the electron density of the carbonyl carbon in such a way as to increase the energy of activation. Louis P. Hammett, in the first edition (1940) of his book *Physical Organic Chemistry*, argued that the effect of the *ortho* substituent:

> . . . can hardly be determined by the interference due to the mere bulk of the substituent implied by the name *steric hindrance*. The effect may better be called a *proximity* effect or an *ortho* effect, the effect of substituents in the α-position of aliphatic compounds appear to be similar and is included in the first term.

Since 'steric hindrance' seemed to have no theoretical basis in mechanistic organic chemistry, it was often relegated to explain phenomena for which no conventional electronic mechanism could be found. One organic textbook author wrote: 'Steric hindrance has become the last refuge of the puzzled organic chemist.'

Undoubtedly a renewed interest and establishment of the legitimacy of the operation of steric factors in organic reactions was stimulated and advanced by H. C. Brown and his students at Purdue University. Some of the concepts introduced by Brown beginning in the 1940s might be briefly summarized. In a reaction: $AB \rightleftharpoons A + B$, where both A and B contain 'bulky' groups, there may be produced some compression of the groups in the compound, A–B. This 'frontal' or 'F' strain is relieved in the dissociation to $A + B$. As an example of this 'F' strain, Brown observed the effect of the position of alkyl substituents in picoline (2-methylpyridine) on the dissociation constant of the complex formed with trimethylboron:

$$[CH_3 C_5 H_4 \overset{+}{N} \cdot B(CH_3)_3 \rightleftharpoons CH_3 C_5 H_4 N + B(CH_3)_3]$$

Brown also proposed a 'back' strain (B-strain), originally to account for the variations of base strength of amines. In the dissociation of the trimethylammonium ion to give trimethylamine and hydrogen ion, the compression of the methyl groups in the ammonium ion (a tetrahedral nitrogen) is relieved in the amine by the expansion of the C–N–C angle. This argument now seems to be even more applicable to the solvolysis of highly branched alkyl halides (i.e. $RCl \rightarrow R^+ + Cl^-$) where the strain is released in the formation of the carbonium ion in the rate-determining step. Recent work of Robert Taft and Edward Arnett on the gas-phase 'basicity' of amines has shown that the solvent plays a major role in determining the orders of basicity in solution.

The importance of steric effects in substitution reactions had also been recognized for some time. The relative inertness of neopentyl halides, $(CH_3)_3CCH_2X$, to solvolysis was pointed out by Frank Whitmore at Pennsylvania State University in 1933. However, Whitmore thought that the stability of neopentyl halides was due to 'the tenacity with which the neopentyl group holds its electrons'. The mechanistic studies of Hughes and Ingold, and M. Polanyi, in the 1930s revealed that the bimolecular nucleophilic attack (S_N2) on alkyl halides was subject to a steric retardation. As the size of the alkyl group near the reactive center increases in bulk, the rate decreases. For example, the relative rates of reaction of alkyl bromides with radioactive bromide ion is found to be:

$$[CH_3CH_2Br > CH_3CH_2CH_2Br > (CH_3)_2CHCH_2Br > (CH_3)_3CCH_2Br$$

$$= 100 : 65 : 3.3 : 0.0015]$$

Since it was observed that both steric and polar (electronic) effects contributed to rate differences, a number of 'physical organic' chemists attempted to formulate equations that could quantitatively distinguish between the two effects. Some of the earlier attempts were made by Karl Kindler (1928) and C. K. Ingold. The failure of the Hammett equation to correlate with the reactivity of *ortho*-substituted benzoic acids was recognized in 1937 by Hammett, who concluded that 'the present correlation of rate and base strengths seems to offer a definite criterion of the magnitude of the steric hindrances'. It was not until the 1950s that Robert Taft succeeded in developing an equation, similar to the Hammett equation, that assigned steric parameters to substituents. A renewed interest in the importance of steric effects was stimulated by the publication in 1956 of the book *Steric Effects in Organic Chemistry*, edited by Melvin Newman.

C.   ATROPOISOMERIC COMPOUNDS

Although some unusual 'space-filling' structures of benzene had been proposed even in the early decades of the 20th century, most chemists thought of benzene as having the shape of a regular hexagon. Biphenyl ($C_6H_5-C_6H_5$) should involve the linking of two planar phenyl rings by a single bond. In 1907, Felix Käufler proposed an alternative structure based on a number of experimental studies. The facile rearrangement of

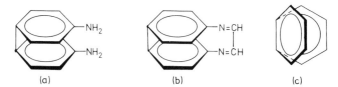

Figure 66. F. Käufler's folded structures of (a) benzidine, (b) the benzidine–glyoxal condensation product, (c) naphthalene.

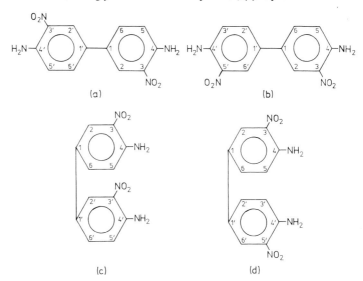

Figure 67. Proposed structures of the 3,3'-dinitrobenzidine 'isomers': (a, b) coplanar (1912), (c, d) Käufler structures (1914).

hydrazobenzene ($C_6 H_5 NHNHC_6 H_5$) to benzidine ($H_2 NC_6 H_4 C_6 H_4 NH_2$) suggested to Käufler that the amino groups in benzidine might still be in close proximity. The formation of a condensation product from the reaction of benzidine and glyoxal was seen as a confirmation of this view. The 'folded' structure of benzidine, and its condensation product on reaction with glyoxal, proposed by Käufler is illustrated in Fig. 66. Käufler also suggested an unusual structure for naphthalene (Fig. 66c).

In 1912, John Cain, Albert Coulthard and Frances Micklethwait isolated two isomers of 3,3'-dinitrobenzidine. The structures of the two isomers that were proposed are illustrated in Fig. 67a, b. The reason for the forbidden rotation about the bond connecting the two phenyl rings was not discussed. Two years later, however, Cain had accumulated additional experimental evidence which he felt made these formulas untenable and proposed that the two isomers should be represented by Käufler structures (Fig. 67c, d). The two isomers were designated as 3,3'- and 3,5'-dinitro-4,4'-diaminobiphenyl, respectively. Following the publication of this article, H. King pointed out that if these were the structures, the 3,5'-dinitro compound should be resolvable into optical isomers. In 1921, James Kenner and W. F. Stubbing at the University of Sheffield reported a resolution of another biphenyl compound, which was designated as the gamma (γ) or 'trans' isomer of 6,6'-dinitrodiphenic acid. The 'trans' compound was then thought to be a stereoisomer of a beta (β) or 'cis' compound prepared in 1903 by Schmidt and Kämpf. Kenner, however, did not feel that the results could only be interpreted in terms of a Käufler structure. Three structures should be considered: (1) formulas in which the phenyl rings were coplanar, (2) the Käufler formula, and (3) those formulas in which the 'two benzene nuclei possess a common axis, but do not lie in the same plane'. In the first case, neither isomer should be resolvable; in the second, only the *trans* isomer should be resolvable; and in the last, both isomers should be resolvable. Although the Käufler structure seemed the most attractive since only one of the isomers had been obtained in an optically active form, Kenner and Stubbings felt more studies were needed before the question could be decided. Within a few years, several optically active biphenyl compounds had been obtained. No *cis/trans* isomers were found, however. By 1926, Kenner was able to show that the dinitrodiphenic acid prepared by Schulze and Kämpf was in fact a structural isomer (2,4'-dinitro-6,6'-dicarboxybiphenyl) of the compound that he had obtained in optically active form earlier. Since the two isomeric dinitrobenzidine compounds reported earlier by Cain had also been shown (in 1920) to be structural isomers, no examples remained of *cis/trans* isomers in the biphenyl series and the Käufler formula now seemed less likely. But if the Käufler formula could not be used, the question was still left unanswered as to how the asymmetry was maintained in the optically active biphenyl compounds. Three groups of investigators simultaneously proposed solutions to this question in a series of communications in the 1926 issues of the magazine *Chemistry and Industry*. All suggested that a rotational barrier existed about the single bond joining the two rings.

   E. E. Turner and R. J. W. Le Fevre (at East London College) assumed that each carbon atom in the ring possessed a certain amount of 'free affinity', which was mutually saturated (in the 'Thiele' sense) for most of the carbons in the rings. The saturation of affinities between the rings (between the 1,1'- and 2,2'-carbons) might help stabilize the coplanar configuration. The introduction of substituents on the 2- and 2'- positions would affect the amount of free affinity available for this stabilization: when substituents were such that they introduced repulsive forces between the carbons, a non-planar configuration of the two rings might be favored, and the compound could be obtained in an optically active form. This proposal does not seem to have received further consideration by other chemists, who were inclined to accept the more mechanical explanations offered by the other investigators. Frank Bell and Joseph Kenyon (at Battersea Polytechnic) were inclined 'to view that the asymmetry may be bound up with the electrical character and size of the substituent groups in the sense that [substituents in the 2,2', 6,6'- positions] may act as obstacles' to the conversion of one enantiomorph into the other. W. H. Mills (at the University of Cambridge) demonstrated with diagrams that when the constituent atoms were represented by spheres, substituents in the 2,2', 6'- positions in 6-chloro-2,2'-dicarboxybiphenyl would prevent the free rotation about the bond joining the rings. These diagrams were presumably based on models that he had constructed himself. (Commercially available models that indicated the correct interatomic and Van der Waals radii were not sold until the 1930s.)

The publication of these papers prompted a number of experimental investigations to determine which substituent groups might be effective in producing the rotational barrier. In the next 15 years, there appeared reports of resolutions (or attempts) of over 50 biphenyl derivatives. The data available from these reports were employed by Roger Adams (at the University of Illinois) to calculate what was termed the 'interference value' of the substituents. Although the magnitude and sign of the interference values correlated with the relative rates of racemization, the correlation was at best only qualitative since the interference values were based only on the distances between the substances and did not take into account the size of the groups. More satisfactory quantitative correlations were obtained by Frank Westheimer at Harvard University in 1946–47. Starting from experimentally available data of the Van der Waals radii of the substituents and the stretching and bending force constants of the bonds in the substituent, the enthalpies of activation for the racemization of several biphenyl compounds could be calculated. The values obtained were found to agree reasonably well with experimental values:

Biphenyl derivative	$\Delta H^{\ddagger}$ (kcal/mole)	
	Calc.	Obs.
2,2'-dibromo-4,4'-dicarboxy	18.2	19.0
2,2'-diiodo-5,5'-dicarboxy	21.4–23.6	21.0
2,2',3,3'-tetraiodo-5,5'-dicarboxy	28.6–33.1	27.3

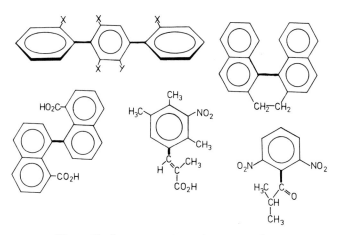

Figure 68. Some atropoisomeric compounds.

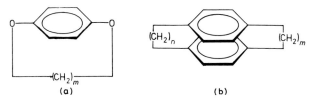

Figure 69. (a) An 'ansa' compound (the compound is optically stable when $m = 8$; racemizes on heating when $m = 9$; unresolvable when $m = 10$). (b) A paracyclophane (the compound is optically stable when $n, m = 2$; racemizes when $n = 4$, $m = 3$; unresolvable when $n, m = 4$).

The biphenyl compounds that were isolated as stable stereoisomers are examples of what are now termed *atropoisomers*. Atropoisomerism is a special case of conformational isomerism in which the isolation of a particular conformational isomer under normal experimental conditions is made possible by a high rotational barrier. A few of the many types of atropoisomeric compounds that have been reported in the literature are shown in Fig. 68.

In the 'ansa compounds' (Fig. 69a) and 'paracyclophanes' (Fig. 69b), when the bridge is small, the phenyl ring (or rings) cannot swivel through the alicyclic ring and the molecules can be isolated in their optically active forms.

In 1956, Melvin S. Newman and D. Lednicer (at Ohio State University) reported the preparation of optically active hexahelicene (Fig. 70). The specific rotation of the compound was found to be unusually large

$([\alpha]_D^{25} = +3700°)$. The terminal rings were twisted out of a coplanar arrangement by molecular crowding.

Figure 70. (+)-Hexahelicene ($C_{26}H_{16}$).

13

The stereochemistry of cyclic compounds: the early history

By the 1870s the only cyclic compounds known to organic chemists were to be found in the aromatic substances, and these consisted for the most part of derivatives of benzene. The few attempts that had been made to synthesize rings containing fewer than six carbon atoms were unsuccessful. The treatment of 1,3-dibromopropane with sodium, for example, was reported to have produced propylene rather than cyclopropane. The lack of success in these syntheses led Victor Meyer to conclude in 1876 that it was unlikely that any such small ring compounds could be prepared.

Plate 13.   Johann Friedrich Wilhelm Adolf von Baeyer (1835–1917)
(Edgar Fahs Smith Collection).

In 1880, William Henry Perkin while in Wislicenus' laboratory translated Meyer's article in order to improve his facility with the language.
When Perkin later moved to Adolf von Baeyer's laboratory in 1882, he
had an opportunity to ask Meyer persónally on a visit there whether he
still held the views expressed earlier. The result of this inquiry was that
Perkin and Meyer spent the evening at the Hofbräuhaus discussing the
problem. When Perkin told him of his intention to prepare such small ring
compounds, Meyer '. . . was much impressed by my enthusiasm but
thought I should be well advised at so early a stage in my career to work
at something more promising and more likely to give positive results'.
Baeyer (Plate 13) was equally discouraging and inquired further how it
was that such compounds had never been found in Nature. (Emil Fischer
not only felt that such compounds could not be synthesized, but even if
synthesized they would likely be produced in such a small quantity and so
unstable as to make it difficult to demonstrate their existence.) The
general acceptance of these views by many organic chemists of the period
probably accounts for the fact that when reports appeared claiming the
synthesis of small ring compounds (cyclopropane by Freund in 1882,
cyclobutane by Markovnikov and Krestonikov in 1881), they were largely
ignored.

In spite of the discouraging comments received by Perkin, he nevertheless designed and tested a synthesis of the cyclobutane ring. The synthetic
scheme involved the reaction of the sodium salt of a β-ketoester with
1,3-dibromopropane (Fig. 71). Perkin expected that the hydrolysis of the
cyclic ketoester (71a) should produce two cyclobutane derivatives
(71c, d). Although the cyclization reaction did produce a product giving
the correct elemental analysis for the cyclic ketoester (71a), the hydrolysis gave a carboxylic acid whose formula corresponded to the pre-

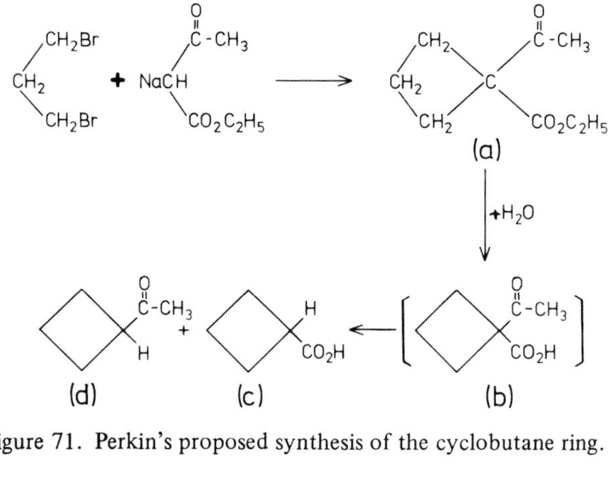

Figure 71. Perkin's proposed synthesis of the cyclobutane ring.

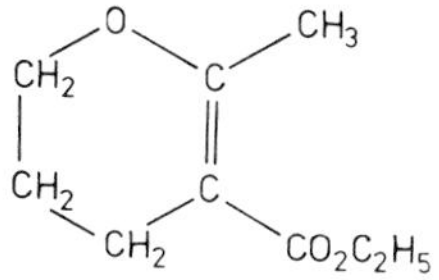

Figure 72.

sumably unstable α-ketocarboxylic acid (71b). The fact that this compound was uncharacteristically stable did not deter Baeyer from sending a preliminary communication to the Bavarian Academy in 1883 and publishing a paper in the *Berichte der Deutschen Chemischen Gesellschaft* shortly thereafter. It was later shown that this was not the product but a cyclic compound (Fig. 72) that arose from a C-alkylation and O-alkylation of an enolate anion. Before this result was known, however, Perkin had run a variation of the reaction which did, in fact, produce the diethyl ester of cyclobutanedicarboxylic acid. Shortly thereafter cyclopropanecarboxylic acid was also prepared by Perkin. Although Rudolf Fittig maintained for a time that this latter compound was an unsaturated acyclic carboxylic acid, $CH_2=CHCH_2CO_2H$, the correctness of the cyclic structure was experimentally confirmed by Perkin. Finally, in 1885, Perkin was able to prepare a derivative of a five-membered ring.

Adolf von Baeyer was concerned with why it was that the three- and four-membered ring compounds were chemically less stable, and more difficult to synthesize, than the six-membered rings. After thinking about this for some time, he proposed that the instability was due to the existence of a 'strain' or 'tension' in the bonds. The theory appeared as an addendum to a longer paper concerned with the preparation and properties of several di- and tri-acetylene compounds. The explosive instability of these acetylene compounds might be thought of as related to the 'tension' of the triple bonds which could be understood in terms of Van't Hoff's concept of the tetrahedral carbon. Perkin later tells us that on two occasions Baeyer invited him into his study and explained his theory with the aid of specially prepared models (these are presumably the modified Kekulé models discussed in Chapter 3). The origin of strain in cyclic compounds was explained as follows:

> The direction of these attractions [between the carbon atoms] can undergo a diversion (from the normal direction of the valencies toward the corners of a tetrahedron) which causes a strain which increases with the size of the diversion. The meaning of this statement can easily be explained if we start from the Kekulé spherical model and assume that the wires, like elastic springs are moveable in all directions. If now, the explanation

that the direction of the attraction always coincides with the direction of the wires is also assumed, a true picture is obtained of the hypothesis outlined in the seventh statement [i.e. that the angle between the valencies is 109° 28']. If now, as can be shown clearly by the use of a model, an attempt is made to join a greater number of carbon atoms without force, that is, in the direction of the tetrahedral axes, or the wires of the models, the result is either a zig-zag line or a ring of five atoms, which is entirely comprehensible since the angles of a regular pentagon, 108°, differ only slightly from the angle 109° 28' which the axes of attraction make with one another. When a larger or smaller ring is formed, the wires must be bent, i.e. there occurs a strain . . .

Baeyer then went on to calculate the angular distortion in the rings. These were summarized in a figure included in the paper (Fig. 73). The double bond was considered a 'two-membered' ring, which is consistent with how Baeyer determined the 'strain' from the Kekulé models he used.

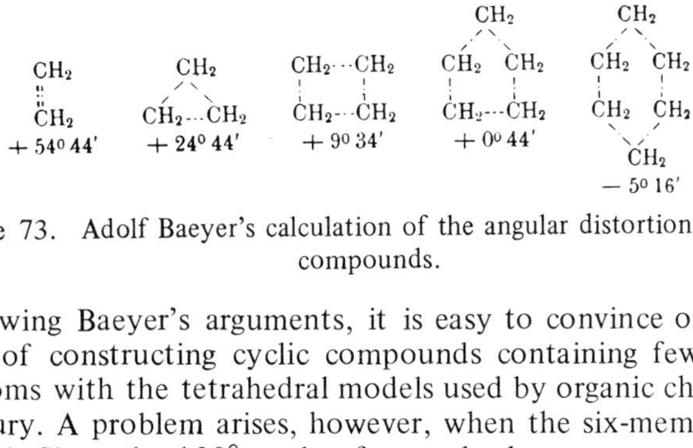

$$
\begin{array}{ccccc}
 & & & CH_2 & CH_2 \\
 & & & /\backslash & /\backslash \\
CH_2 & CH_2 & CH_2\cdots CH_2 & CH_2\;\;CH_2 & CH_2\;\;CH_2 \\
\| & /\backslash & | & | & | & | \\
CH_2 & CH_2\cdots CH_2 & CH_2\cdots CH_2 & CH_2\cdots CH_2 & CH_2\;\;CH_2 \\
+54^\circ44' & +24^\circ44' & +9^\circ34' & +0^\circ44' & \backslash/ \\
 & & & & CH_2 \\
 & & & & -5^\circ16'
\end{array}
$$

Figure 73. Adolf Baeyer's calculation of the angular distortion in cyclic compounds.

In following Baeyer's arguments, it is easy to convince oneself of the difficulty of constructing cyclic compounds containing fewer than five carbon atoms with the tetrahedral models used by organic chemists in the 20th century. A problem arises, however, when the six-membered ring is constructed. Since the 120° angle of a regular hexagon exceeds the tetrahedral angle, Baeyer assigned a 'negative' strain to the cyclohexane ring. When, however, cyclohexane is constructed with tetrahedral models that use springs or wires to connect the atoms, it is apparent that the ring is not planar *or* strained. That Baeyer did not comment on this has struck some chemists as rather curious. Since, however, it was primarily Baeyer who had demonstrated experimentally the relationship between cyclohexane and benzene, it might be assumed that by that date he must have felt that the hexagonal shape of benzene must be preserved in cyclohexane.

As was mentioned earlier (Chapter 3), Baeyer probably modified the Kekulé models in the 1880s in such a way that the valency wires in double

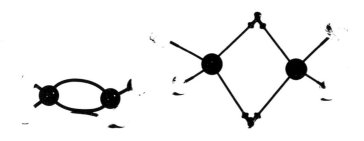

Figure 74.   Models of ethylene, $H_2C=CH_2$: (left) 'spring-bond' model;
(right) Kekulé–Von Baeyer model.

and triple bonds were connected with a flexible joint rather than a wire
loop. When these models are used to construct a carbon–carbon double
bond, as for example in the ethylene model shown on the right in Fig. 74,
the wires are not bent, as is the case in the 'spring-bond' models shown on
the left. The strain attributed to the double or triple bond must be considered in conceptual terms as arising from the fact that the valency forces
(as wires) are not linearly directed toward each other to produce the
maximum attractive force. When cyclic compounds are constructed with
these models (remembering that *two* wires are used to form a bond), the

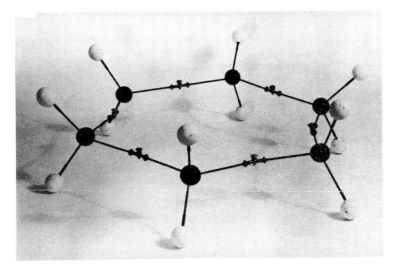

Figure 75.   Cyclohexane constructed with Kekulé–Von Baeyer models.

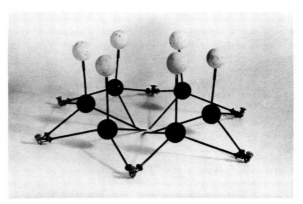

Figure 76. Model of Adolf von Baeyer's structure of benzene.

angle between the wires increases as the ring becomes larger. In a six-membered ring the wires are joined inside the ring (Fig. 75), producing a 'negative' strain.

One of Baeyer's earlier proposals for the structure of benzene was similar to the Koerner structure (Fig. 15, Chapter 3) except that the six carbons were coplanar and all six hydrogens were placed on one side of the ring (Fig. 76). The positioning of the hydrogens in this way was justified by the observation that the hydrogenation of phthalic acid gave *cis*-hexahydrophthalic acid, and not the *trans* isomer. When the hydrogenation of benzene is illustrated with models, the carbon atoms maintain their coplanarity (Fig. 76 + 3H$_2$ → Fig. 75). Furthermore, there are observed two sets of hydrogens, six above and six below the plane of carbon atoms. With this model, Baeyer was able to explain how it was

Maleic anhydride   Fumaric anhydride

Anhydride of   Anhydride of
cyclohexane − *cis*-1,2-dicarboxylic acid   cyclohexane − *trans*-1,2-dicarboxylic acid
('Malenoid' hexahydrophthalic acid)   ('Fumaroid' hexahydrophthalic acid)

Figure 77.

possible for a cyclic anhydride to be formed from both *cis-* and *trans-*cyclohexane-1,2-dicarboxylic acid (Fig. 77). As he observed in 1890, an examination of the models of these two isomers reveals that the two carboxyl groups are close enough (at a 'dihedral' angle of about 109°) in both to allow the formation of the cyclic anhydride. The two carboxylic groups in fumaric acid are too far apart (180°) to permit the formation of a cyclic anhydride. Since the 1890 paper did not include any perspective drawings, Baeyer's speculation apparently escaped the attention of most organic chemists (and historians). It was, in fact, Baeyer who pointed out the analogy of the *cis* and *trans* isomers of 1,2-, 1,3- and 1,4-hexahydrophthalic acid to the geometry of maleic acid and fumaric acid. Originally he referred to the isomers as the 'malenoid' and 'fumaroid' forms. The simple hexagonal picture of the six-membered ring used by Baeyer and most organic chemists thereafter, obscured the more subtle stereochemical relationships in cyclohexane derivatives that had been considered earlier by Baeyer. In subsequent decades, most organic chemists adopted a planar structure for representing the six-membered ring as the most convenient means of illustrating the structures of 'geometric' isomers.

In 1890, Victor Meyer, commenting on the merits and deficiencies of the strain theory, suggested that the experimentally observed greater stability of the six-membered ring, compared to the five-membered ring, might be found in the greater symmetry of the former. He also noted that the angle strain, although larger than for cyclopentane, was still quite small. He also thought that the presence of strain in a triple bond seemed inconsistent with the observation that acetylene was the only hydrocarbon formed when a carbon arc was sparked in an atmosphere of hydrogen.

Baeyer had reported that an isomerization of the 'malenoid' to 'fumaroid' form of hexahydrophthalic acid took place even at room temperature, an isomerization which could not be explained in terms of the mechanisms proposed by Wislicenus for the isomerization of maleic acid to fumaric acid (Chapter 6). In a paper published in 1890, Hermann Sachse sought to explain this rather facile isomerization in terms of nonplanar structures. He felt an alternative was needed since 'the atoms or groups joined to the four valencies (of the carbon atom) cannot simply change places with each other'. Sachse pointed out that if the model of cyclohexane was constructed with tetrahedral carbons, two strain-free conformations (the term 'configurations' was used by Sachse at that time) of cyclohexane were possible: one, a rigid, symmetrical form (later known as the 'chair' or 'Z' form); and the second, a flexible, unsymmetrical form (the 'boat', 'tub' or 'c' form). Sachse showed how models of the two forms could be constructed using cardboard tetrahedra, but no three-dimensional illustrations were included. The 'normal' configuration of cyclohexane was considered to be the symmetrical ('chair') form. (He later attempted to justify this choice by means of mathematical calculations.) The structure was also related to a space-filling model of benzene he had proposed two years earlier. In the 'fumaroid' form of hexahydrophthalic acid, Sachse placed the two carboxyl groups in what we would now designate as axial positions; in the 'malenoid' isomer, the carboxyl groups would be in 'equatorial' positions. The isomerization is explained in terms of interconversions of the two chair conformations (fumaroid = a, a → malenoid = e, e). Sachse also recognized that a third isomer (which would correspond to the carboxyls in axial and equatorial positions) was also possible, but was unknown. [No isomerization is produced in the interconversions of the chair conformations (i.e. a, e → e, a) of this isomer.]

The comparative ease with which the interconversion of the isomers of hexahydrophthalic acid took place was attributed by Sachse to the repulsive action of the carboxyl groups. In the absence of substituents, however, Sachse thought that a higher temperature might be required to produce the interconversion. It was, therefore, possible that two isomers of a monosubstituted cyclohexane derivative might be capable of isolation if the separation were carried out at room temperature. Since later attempts by chemists to isolate such isomeric forms were unsuccessful, no experimental evidence was available for some time that supported Sachse's idea of a multiplanar conformation of cyclohexane. (It was not until 1966 that F. R. Jensen and C. H. Bushweller were able to demonstrate a low temperature separation of the axial and equatorial forms of chlorocyclohexane.)

Although the planar form of cyclohexane was widely used by chemists to illustrate the structure and reactions of cyclohexane derivatives, a surprising number of experimental investigations were undertaken in the latter part of the 19th and early 20th century to test the validity of the Baeyer strain theory and Sachse's idea of multiplanar rings. In 1891, F. Stohman and C. Kleber published data concerning the heats of formation of a number of cyclic compounds. These data were frequently cited as supportive of the strain theory, even though the slightly greater stability of the six-membered ring over the five-membered ring was apparent. Suggestions that the five-membered ring might be more stable were usually based on the results of rearrangement reactions, such as was observed in the conversion of pinene to pinene hydrochloride and Demjananow rearrangement of cyclobutylmethylamine to cyclopentanol. Other rearrangement reactions, however, indicated that the two rings were of about the same stability. Experimental data about larger rings were not very extensive in this period; and it was, therefore, generally assumed that these cyclic compounds should be rather unstable. (By 1902, however,

Perkin was not sure that the strain theory was useful in understanding the stability of large rings having negative angles of torsion.)

In 1899, Werner and Conrad reported that only the *trans* isomer of hexahydrophthalic acid could be resolved into its enantiomers. This result, of course, was compatible with the planar-ring structure but, in fact, the investigation was originally undertaken in order to test Sachse's concept of the strainless ring. Werner and Conrad apparently did not realize that the same result would be obtained if the various conformations were easily interconvertible. Most chemists subsequently cited the report as evidence against the multiplanar form. In 1905, Ossian Aschan pointed out that when the cyclohexane molecule was constructed with tetrahedral models, the different 'phases' of cyclohexane were easily interconvertible, and therefore were probably not capable of isolation. Since the planar configuration of cyclohexane corresponds to an intermediate phase in the interconversion of the multiplanar forms, it was therefore more useful to consider the structure of cyclohexane in terms of this intermediate, planar form. This view persisted in the literature for some 30 years. For example, in the chapter entitled 'Stereoisomerism' (written by Shriner, Adams and Marvel) in the 1938 edition of Gilman's textbook of organic chemistry is found the following statement:

> However, no isomeric forms of cyclohexane have been isolated up to the present time. It seems probable that there is an equilibrium between the two forms, and that the two models vibrate from one to the other so rapidly that the net average result is a planar molecule.

The publications of Ernest W. M. Mohr (at Heidelberg) in 1915—18 rekindled interest in the possibility of the multiplanar forms of cyclic systems. Mohr first of all pointed out that the existence of numerous

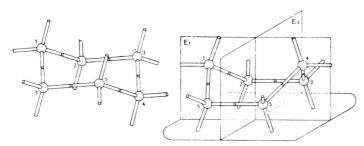

Figure 78.   Ernest Mohr's illustration of the chair and boat forms of cyclohexane.

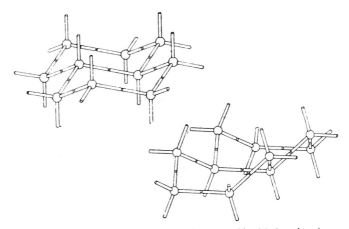

Figure 79.   Ernest Mohr's illustration of the *cis* 'double-boat' and *trans* 'double-chair' conformation of decalin.

bicyclic and tricyclic substances (such as camphor and adamantane) were inexplicable on the basis of planar structures. Such molecules were easily constructed with tetrahedral models. The impact of Mohr's papers was probably greater than that of Sachse's since he illustrated his articles with numerous three-dimensional drawings, which might have been based on Kekulé models (Fig. 78). The chair and boat forms of cyclohexane were both shown. Mohr argued that the resistance needed to convert one conformation to another was probably much smaller than Sachse believed and that normal thermal collisions would be adequate to bring about the interconversions of the various multiplanar forms. The stability of the chair form seemed consistent with the structure of diamond which had been determined by X-ray structural analysis by W. H. and W. L. Bragg in 1913. However, in the fused-ring compounds such as decahydro-naphthalene (decalin) this interconversion was not possible and Mohr therefore predicted that there should exist two strain-free *cis/trans* isomers (Fig. 79). Mohr's prediction received experimental support in 1923 when A. Windaus, W. Hückel and C. Reverez at Göttingen reported the synthesis of the cyclic anhydrides of *cis*- and *trans*-hexahydrohomo-phthalic acid (Fig. 80). It is difficult to see how the latter compound

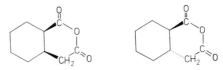

Figure 80.   *Cis*- and *trans*-hexahydrohomophthalic anhydride.

could be formed if the ring were planar, but is understandable in terms of the close proximity of the carboxyl groups when the ring is in a chair conformation. The fact that Baeyer had observed the formation of similar cyclic anhydrides (Fig. 77) some 25 years earlier had apparently been forgotten. Since Hückel's conclusions were not accepted (apparently on the basis that nothing was known about the direction of the oxygen valencies), he undertook the preparation of the completely carbocyclic systems whose existence had been predicted by Mohr. In the same year Hückel reported the synthesis of two isomers of decahydro-β-naphthol (2-hydroxydecalin) and the corresponding ketones. Although these syntheses provided persuasive evidence for the existence of a multiplanar cyclohexane ring, some chemists still argued that the bicyclic compounds were unique cases in which the multiplanar forms had a special stability which might not be found in the simpler monocyclic systems. It would seem that it should not have been difficult to obtain experimental information concerning the conformation of the cyclohexane ring; yet those physical and chemical studies that were reported in the next two decades produced ambiguous and contradictory results. For example, in the discussion that follows the report of S. B. Hendricks and C. Bilicke in 1926 on the X-ray crystallographic study of β-benzene hexachloride, it was proposed that the carbon atoms were coplanar. However, in the diagrams (based on the X-ray data) that are included in the article, it is clear that the carbon atoms describe a chair conformation.

The persistent reports of the existence of an excessive number of 'conformational' isomers made it difficult to accept either the multiplanar or planar structures. In 1925, for example, W. A. Wightman thought that the report of a third isomer of decalin rendered Mohr's ideas invalid. Yet a year later he was suggesting that the planar structure could not explain the report of three isomers of bicyclohexane ($C_6H_{11}C_6H_{11}$). The isomerism could be explained in terms of non-interchangeable multiplanar structures. It might be mentioned that often the evidence used to support the isolation of isomeric compounds was meagre by modern standards. Sometimes the existence of two isomers was based upon the isolation of two distillation fractions of different refractive indexes. Since these reports were so inconclusive, Wightman set about to devise an unambiguous test for the existence of a multiplanar form. Briefly summarized, the method involved a series of chemical changes (Fig. 81) of a 1,1-disubstituted cyclohexane which produced an exchange of the substituents

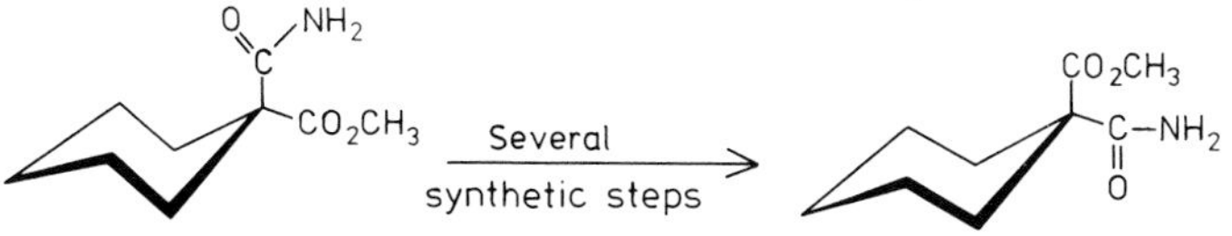

Figure 81.

without breaking any bonds to the ring. Since the two substances proved to be identical (as judged from melting point behavior), Wightman concluded that 'the rigid [multiplanar] form of cyclohexane does not exist as a static modification; the possibility of dynamic equilibrium is not, however, excluded'. This report did not discourage further attempts to isolate such isomers. In fact, in the 1930s a number of reports were made of compounds having an excess of isomers (later shown to be structural isomers). The search for stable 'conformational' isomers in this period also seemed compatible with information provided by the molecular models that were used. The models often suggested a greater stability than was the actual case. For example, in 1938 Carl Marvel and M. A. Glass attempted the resolution of a derivative of cyclononane since models of cyclononane showed that 'two arrangements were possible and that one of the models was a non-superimposable image of the other'. The models used were commercial models based on those first described in 1934 by H. A. Stuart at the Institute of Physics at Königsberg. In this 'space-filling' model, each atom occupied a roughly spherical volume whose size was related to the Van der Waals radius. The molecules constructed with these models are often found to be rather inflexible due to the collision of non-bonded atoms, even though no such 'severe' interaction is observed experimentally. It does not appear that availability of these

TABLE II

Calculations of the internal (α) and exterior (β) angles in cyclic compounds according to the Thorpe—Ingold 'valency-deflection' hypothesis

Ring size	α	β
3	60°	117°
4	90°	113°
5	108°	109°
6	120°	107°
7	128°	105°
8	135°	103°

kinds of models had any detectable impact upon the earlier development of conformational analysis. It should be pointed out, however, that no rotational barrier is observed in the ethane molecule constructed with the Stuart models.

The chemists' view of a strained planar cyclohexane ring might also have been given added credibility following the reports of a series of investigations carried out primarily by J. F. Thorpe and C. K. Ingold at University College in London from 1915 to 1930 in support of what was called a 'valency-deflection' hypothesis. Briefly summarized: in a non-cyclic compound such as propane, it was proposed that because of the larger atomic volume of carbon relative to hydrogen, the C–C–C valency angle should be larger than the normal tetrahedral angle (about $115.3°$). In cyclic compounds, since the interior C–C–C angle was controlled by the size of the (planar) ring, the external angle, β (Table II), would vary with the ring size. A consequence of the decreasing angle external to the ring should be to facilitate ring closure reactions. Experimental support for the hypothesis was derived from determining the yields of product or the magnitude of equilibrium constants found for several cyclization reactions, one example of which is illustrated in Fig. 82. It seemed clear that the data on the six-membered ring did not fit into the progression suggested in Table II. The evidence, in fact, pointed to the existence of cyclohexane in a multiplanar form. Even as late as 1931, Thorpe persisted in his belief that the cyclohexane ring must be planar:

> There is then considerable evidence in favor of a strained cyclo-hexane structure both from the chemical and physical side and none whatever in favor of the strainless multiplanar ring, and it would seem to follow that, when one carbon atom of a six-membered ring is attached to two side chains, the strain set up by the enlargement of the tetrahedral angle from $109.5°$ to $120°$ causes a deflexion of the normal angle made by the other two valencies, the six-membered ring remaining uniplanar.

It is curious that at the same time Thorpe did allow that the seven- and eight-membered rings probably existed in multiplanar forms, presumably because these rings were flexible enough to relieve the strain found in the planar form.

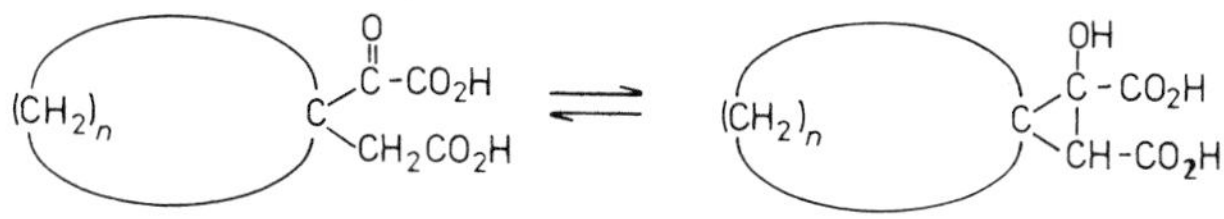

Figure 82.

By this time, the applicability of the strain theory to larger rings was seriously questioned by the first reports of the isolation of macrocyclic compounds from natural products. Although it was known in the early part of the 20th century that the substances responsible for the fragrance of extracts from the Asiatic musk deer and the African civet cat were ketones having a high molecular weight, it was not until 1926 that Leopold Ružička in Zürich demonstrated that muskone and civetone were cyclic ketones containing 15 and 17 carbon atoms. Max Stoll, who was undertaking his doctoral work at the Federal Polytechnic School at that time, recalled Ružička's saying:

> Now, we are faced with a terrible dilemma. A long time ago, Adolf von Baeyer demonstrated that rings of carbon atoms in excess of six in number are difficult of formation and that the difficulty increases with an increase in the number of carbons. And here we have a ring with 17 carbon atoms. How shall we ever succeed in synthesizing it? And how may the industry, for which I work, ever realize a profit from this discovery and produce this compound in good yield?

This comment also reveals another aspect of the interpretation of the strain theory that has not been discussed: namely, that it was generally assumed that there should be a correlation between the stability of a ring and the ease with which it could be synthesized. Stoll was given the task of synthesizing these macrocyclic rings and after many trials, was successful. As the synthesis of macrocyclic rings was examined more systematically, it was found that the yields did not decrease in a regular way as the size of the ring was increased. Ružička considered that the variation in the yields was the resultant of the operation of two factors: the first related to the probability of finding the two ends of the chain in a position for cyclization, and secondly, to a strain factor, which is significant only for the smaller rings and levels off for rings containing more than five carbon atoms. It was noted, however, that the yields of 'medium' rings (C_8-C_{12}) were lower than was predicted on the basis of the operation of these factors alone. In 1930, M. Stoll and G. Stoll-Compte suggested this was due to the presence of steric interferences produced by some of the hydrogens internal to the ring, which is most pronounced in the C_{11} and C_{12} rings. More recent conformational studies have shown that the situation is even more complicated than is indicated here. (A review of some of the problems associated with the synthesis of medium-ring compounds is discussed in Prelog's 1950 paper reprinted in Chapter 1.)

Although these syntheses dramatically illustrated the non-planarity of macrocyclic rings, it is interesting to note that because Thorpe and

Ingold's views were still widely accepted even in 1935, Ružička considered that the six-membered ring might be planar:

> But the circumstances existing in the 6-ring are not the same as in the higher-membered rings, the stability of which can only be explained by the acceptance of Sachse's theory . . .

Some of the most perceptive observations and experiments that provided evidence in support of the existence of a strainless, non-planar form of the six- and seven-membered ring are to be found in the investigations undertaken by Jacob Böeseken and several of his students in Delft, Holland, in the early 1920s. These studies might well have been considered as the origin of conformational analysis, if it were not for the fact that later chemists were for the most part unaware or unappreciative of the research.

Böeseken was interested in examining the cause of the increase of acidity of boric acid solutions to which had been added polyhydroxy compounds. Magnanini, who had first observed this phenomenon in the 1890s, had been content to suggest that some sort of relationship should exist between the constitution of these compounds and their effect on boric acid. No further suggestions were made until 1908, when Van't Hoff proposed that the vicinal diols reacted with the boric acid to form a cyclic compound (having a greater acidity than boric acid):

$$
\begin{array}{c}
-\overset{|}{C}-OH \\
\overset{|}{\underset{|}{}} \\
-\overset{|}{C}-OH \\
|
\end{array}
\;+\; H_3BO_3 \;\longrightarrow\;
\begin{array}{c}
-\overset{|}{C}-O \\
\overset{|}{\underset{|}{}} \quad\!\!\! \searrow B-OH \;+\; 2H_2O \\
-\overset{|}{C}-O \nearrow \\
|
\end{array}
$$

In 1913, Böeseken undertook the measurement of the conductivity of boric acid solutions containing a variety of polyhydroxy compounds. Böeseken suggested that the two hydroxyl groups should not only be on adjacent carbon atoms, but in the same plane. Since 1,2-ethanediol did not produce an increase in conductivity, he assumed that the two hydroxyls could exert a 'natural mutual repulsion' and therefore rotate away from each other to a position unfavorable for the complex-formation with boric acid. In glycerol (which produces a small but measurable conductivity increase) the chance of finding two of the three hydroxyl groups in a proper orientation was greater and therefore there is a greater probability for forming the cyclic borate ester. The experimental quantity 'delta' (Δ) (that is, the difference in conductivity of a mixture of the boric acid solution and the substance, and that of the boric acid solution alone) was regarded as a measure of 'the relative position of the two hydroxyls in space'.

In 1919, one of Böeseken's students, Chr. van Loon, found that only the *cis* but not the *trans* isomers of rigid cyclic diols (such as cyclopentane-1,2-diol) had high delta values. In addition, he also found that a cyclic ketal could be prepared by the reaction of the *cis* diols with acetone in the presence of HCl. A year later, J. van Griffen found, rather surprisingly, that neither *cis*- nor *trans*-cyclohexane-1,2-diol gave a positive value of delta, although the ketal could be prepared from the *cis* diol. Griffen and Böeseken explained the boric acid results by suggesting that 'the cyclohexane ring seemed to possess a certain flexibility which allowed the adjacent OH groups to obey to a certain extent their "natural repulsion" and thus no longer remain fixed in one and the same plane'. When another Böeseken student, H. G. Derx, examined the behavior of cycloheptane-1,2-diol shortly thereafter, he found, even more surprisingly, that both the *cis* and *trans* isomers produced a large increase in delta, although the *cis* isomer gave the larger increase. Furthermore, the acetone ketals could be prepared from both isomers. Derx (but not Böeseken) was aware of Sachse's publications and saw their applicability to the problem. Derx had special tetrahedral carbon models prepared in order to study the various conformations of the diols. With these models, he could demonstrate that the hydroxyl groups were not in a coplanar arrangement in the chair form of cyclohexane. (In the *cis* diol, the hydroxyls were in axial-equatorial positions; in the *trans* diol, in diequatorial positions – in modern terminology.) It was thought that the formation of the ketal from the *cis* diol was because of an interconversion to the boat form where the two hydroxyl groups were coplanar. In aqueous solution this interconversion presumably did not occur to any great degree. (The differential behavior of the *cis*- and *trans*-cyclohexane-1,2-diol toward acetone in ketal formation was only understood much later and was explained in terms of a 'flattening' of the chair conformation which brought the *cis* hydroxyls (a, e) closer together, while the *trans* hydroxyls (e, e) were pried further apart.)

When the seven-membered ring was constructed with models, it was clear that the greater flexibility of the ring made it easier for the hydroxyl groups in both the *cis* and *trans* diols to form the coplanar configuration thought necessary for reaction with the boric acid or acetone. Many of the chemists who were aware of this work had difficulty in understanding why the 'natural repulsion' of the hydroxyl groups in the flexible seven-membered ring would not be more likely to produce a conformation in which the hydroxyl groups were in a position less favorable for reaction than was possible in the rigid six-membered ring.

In 1923–24, P. H. Hermans, working in the same laboratory, extended these studies in two important ways. First of all, he was able actually to isolate several of the 1 : 1 diol–boric acid complexes (the borate esters) and show that they had about the same acidity as boric acid itself, thereby demonstrating that they could not be responsible for the increase in conductivity of the solutions as Van't Hoff had supposed earlier. He then succeeded in isolating the complexes that were responsible. These were shown to be formed from the reaction of two moles of the diol with one mole of boric acid:

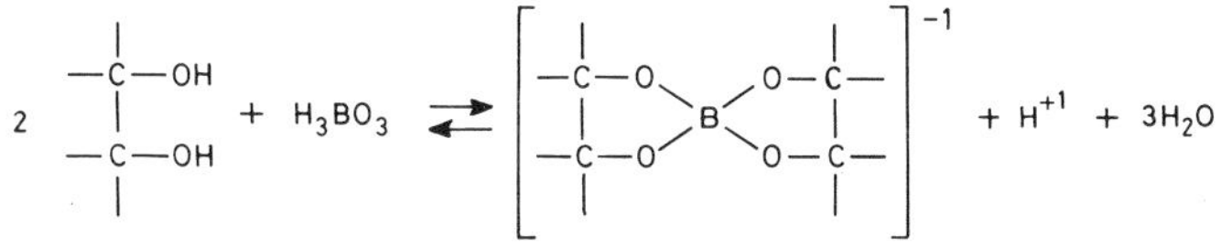

Hermans also predicted that these complexes would have a spirane-type structure and therefore should be resolved into optical isomers. This was demonstrated experimentally by Meulenhoff in 1924.

Secondly, Hermans obtained some precise measurements of the rate and equilibrium of formation (and hydrolysis) of the ketals formed on the reaction of the diols with acetone. The magnitude of the equilibrium constant was shown to correlate with the delta values. Where the conductivity measurements had shown no complex formation for *cis*-cyclohexane-1,2-diol, a small but measurable value of the equilibrium constant ($K_{18}° = 0.18$) for the ketal formation could be determined. (The significance of Hermans' related studies on acyclic diols is discussed in the next chapter.) In order to ascertain what might be the most favorable geometry of the cyclohexane ring, Hermans and J. Berk devised a mathematical approach similar to that used by Barton some 20 years later. The relative potential energies of the various conformations were calculated using the Derx models by considering all of the non-bonded interactions between hydrogens and hydroxyl groups. The bond lengths used in the models were derived from the Van der Waals equation. The calculations demonstrated the greater stability of the 'chair' over the 'boat' form. (It should be mentioned, however, that Barton's calculations were based on Pitzer's earlier studies which showed the preference of staggered over eclipsed conformations in acyclic systems such as ethane.) Although the mathematics required to produce a similar analysis of the seven-membered ring proved too complex, it could be shown with the models that the interconversions between the various conformations were often hindered by the collisions between the 1,4-hydrogens within the ring, an idea that was apparently proposed independently by the Stolls in 1930 to explain the difficulty of preparing medium-sized rings.

It is difficult to understand why the results of the 'Delft' school were not better known. Barton attributed the interest in his work in the 1950s to the fact that there were a large number of chemists involved in the study of steroidal compounds, whose chemical and physical properties could only be properly understood in terms of the conformation of the molecules. In the early 1920s, however, there were few organic chemists who appreciated or were interested in understanding the more subtle relationships between molecular conformation (as distinguished from configurations) and chemical behavior. Physical chemical studies were of interest to organic chemists only in about the 1940s, when we see the rise of what is sometimes referred to as physical-organic chemistry.

14

The origins and development of conformational analysis

A. ACYCLIC COMPOUNDS

To demonstrate a relationship between the conformation of a molecule and its physical or chemical properties requires experimental techniques of greater sensitivity than were employed in the latter part of the 19th and early 20th century. The studies of acyclic systems undertaken in this period in general provided additional support for the belief that the rotation about single bonds was unhindered. The major exception to this was found in a special group of substances, the biphenyls, where the rotation about the bond joining the two phenyl rings could be hindered by large 'bulky' substituents (Chapter 12).

By the 1930s, however, the results of an increasing number of spectroscopic studies and other physical property measurements could only be understood by supposing that molecules had preferred conformations. This conclusion was not immediately appreciated by, or of general interest to, organic chemists. First of all, the idea of a preferred molecular conformation was not supported by the isolation of any of the predicted stereoisomers. Secondly, and probably of greater importance, no 'useful' chemical information could be gained by considering molecular conformations, especially if there was a rapid interconversion between the various conformers. The application of a physical-chemical approach to the study of organic reactions involved only a few organic chemists in this period.

It is therefore not so surprising to find that the experimental studies of P. H. Hermans at Delft in the 1920s were so little known. Were it not for this, the origins of conformational analysis could be found in the results of Hermans' careful quantitative studies of the reactions of acyclic diols, for in this work he was able to demonstrate the more subtle relationships between molecular conformation and chemical reactivity. Hermans hoped that a rate study of the reaction of the diastereoisomeric

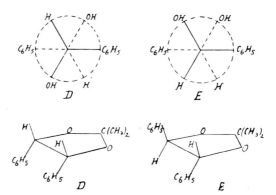

Figure 83.   Perspective formulas used by P. H. Hermans to illustrate the
conformations of the diastereomeric hydrobenzoins (top) and their cyclic
acetone-ketals (bottom).

hydrobenzoins (1,1-diphenylethane-1,2-diol) with acetone to form the
cyclic ketals (D and E at the bottom of Fig. 83) would provide a direct
measure of steric influences in the reaction. The 20-fold increase in rate
of the *racemic* (E) over the *meso* compound (D) was interpreted with
respect to the projection formulas (Fig. 83) included in his dissertation
thesis and an article published in the *Zeitschrift für physikalischen Chemie*
in 1924:

> The configurations with a minimum of potential energy will be
> those where the phenyl groups lie at the maximum distance
> from each other representing a projection in the direction of the
> central C–C bond. The energy required to bring the two OH-
> groups closely together (into the 'favorable' position) will be
> much smaller in the case of the racemic isomer (E) than that of
> the inactive isomer (D), because in the latter it is required also
> to bring the phenyl groups closely together.

The rate difference was shown to be related to the fraction of molecules
in the favorable conformation – an analysis that predates by some 30
years the equations developed by Winstein, Holness and Eliel, Lukach.
The entropies and enthalpies of activation for reaction were also determined. The difference in the energy of activation of the two diastereoisomers was considered to be related to the steric factors involved.

Although a continuation of these kinds of studies might have led to an
earlier establishment of the study of conformational analysis, the relationship between the conformation of a molecule and its reactivity was only
gradually perceived by organic chemists following the publication of the

results of a number of spectroscopic studies and other physical property determinations.

In 1929, Arnold Weissberger at Leipzig determined the dipole moments of a number of p,p'-derivatives of biphenyl and concluded that the results indicated that biphenyl did not exist in a 'folded' conformation (Chapter 12). (A few years earlier, however, he had interpreted the observation of a dipole moment for p,p'-diaminobiphenyl as indicating that biphenyl *did* exist in such a conformation.) In the same year, J. Clark and L. W. Pickett, at the University of Illinois, interpreted their X-ray studies of $2,2',6,6'$-substituted derivatives of biphenyl as indicating that the phenyl rings were not coplanar.

By 1930, dipole moment measurements had been reported for a number of acyclic compounds and these data provided the first physical-chemical evidence, in compounds other than the biphenyls, to suggest that rotation about the carbon–carbon single bond might be hindered. This was indicated, for example, in the dipole moment measurements of the diastereomers of stilbene dichloride and diethyl tartrate (determined by A. Weissberger and K. L. Wolf respectively) (Table III).

TABLE III

Dipole moments of stereoisomers of stilbene dichloride and diethyl tartrate

Diastereomer	$C_6H_5CH(Cl)CH(Cl)C_6H_5$	$C_2H_5O_2CCH(OH)CH(OH)CO_2C_2H_5$
meso-	1.27D	3.66D
dl-	2.75D	3.12D

The diastereomers should not show a difference in the dipole moment if there were free rotation about the central carbon–carbon bond. These and other studies gradually confirmed the existence of a rotational barrier about single bonds. Speculations as to the cause of the barrier initially involved the assumption of some sort of electrostatic interaction between the non-bonded groups.

In the earlier literature in which are found discussions of which molecular conformation might be the least (or most) stable, only the eclipsed conformations (and not the staggered) were thought to be important. (Only when the non-bonded groups were rather 'bulky' was the staggered conformation preferred.) In 1,2-dichloroethane, for example, what was referred to as the '*cis*' eclipsed form was considered to be less stable than the '*trans*' eclipsed form because of the repulsions between the polar chlorine atoms (Fig. 84). The observation that the dipole moment of the

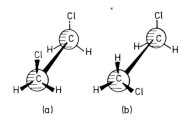

Figure 84.   Eclipsed conformations of 1,2-dichloroethane: (a) the '*cis*'
form, (b) the '*trans*' form.

compound was larger at higher temperatures was also thought to be consistent with this interpretation. Raman spectral studies also suggested that many simple ethane derivatives had preferred conformations. S. Mizushima (at the University of Tokyo) observed that there were more lines observed in the Raman spectra of these compounds than would be predicted for the molecule existing in only one particular conformation. The number of lines was reduced, however, when the spectra were obtained at lower temperatures (which would favor a particular conformation). For 1,2-dichloroethane Mizushima thought that the '*trans*' form would be the most stable conformation.

No rotational barrier could be inferred from the early physical-chemical measurements of ethane itself. The eclipsed conformation was sometimes thought to have a greater stability because of the 'attractions' between the non-bonded hydrogen atoms. In 1932, Henry Eyring (then at Princeton University) calculated that, on the grounds of such non-bonded interactions, the rotational barrier in ethane should only amount to about 0.3 kcal/mole. This estimate was apparently confirmed by the specific heat measurements reported by Eucken and Weigert in 1933. In 1936, however, J. D. Kemp and Kenneth Pitzer at the University of California in Berkeley, argued that the statistical mechanical calculations of the enthalpy and entropy of ethane would not agree with the experimental values unless it was assumed that there was a barrier to rotation about the carbon—carbon bond of about 3 kcal/mole. When the specific heat of ethane was redetermined in 1938 by Kistiakowsky, Lacher and Still, a rotational barrier of about 2.75 kcal/mole was observed. The establishment of the ethane barrier has been considered by some as marking the highlight of the history of the conformational analysis of acyclic systems.

At this point in the discussion, it might be useful to talk more about what is meant by the term 'conformation'. Generally the term is used to describe any of the infinite number of momentary arrangements of the atoms in space that result from rotations about single bonds. The variations in the meaning of this word that have appeared are discussed in a number of standard books concerned with conformational analysis. The

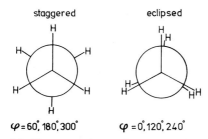

Figure 85. The two major conformations of ethane.

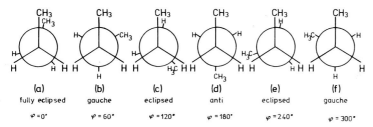

Figure 86. Butane conformations.

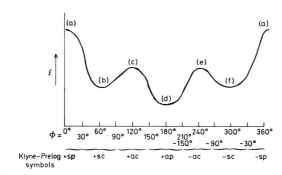

Figure 87. Potential energy diagram for butane conformations.

word 'conformation' was first introduced to chemists in 1929 by H. N. Haworth to describe the chair and boat forms of the pyranose ring. A rather detailed terminology has evolved since the 1930s to describe the conformations of acyclic compounds. This terminology is briefly illustrated with reference to ethane and butane, whose conformations will be illustrated by means of Newman projections. (These projection formulas were proposed by M. S. Newman in 1955 and are almost identical in concept to those used by Hermans and Bischoff much earlier.) When we

TABLE IV

Klyne—Prelog conformational terminology

Angle of torsion	Designation	Symbol
$- 30°$ to $+ 30°$	$\pm$ syn-periplanar	$\pm$ sp
$+ 30°$ to $+ 90°$	$+$ syn-clinal	$+$ sc
$+ 90°$ to $+ 150°$	$+$ anti-clinal	$+$ sc
$+ 150°$ to $210°$ (or $-150°$)	$\pm$ anti-periplanar	$\pm$ ap
$- 30°$ to $- 90°$	$-$ syn-clinal	$-$ sc
$- 90°$ to $-150°$	$-$ anti-clinal	$-$ ac

look along the C–C axis of ethane (Fig. 85), the two major conformations are considered in terms of the angle of rotation, which is denoted variously as the 'dihedral angle' (ϕ), 'angle of torsion' (τ) or 'azimuthal angle' (θ). (The formulas do not actually show complete eclipsing so as to allow for the writing of the groups.)

In butane there are two different staggered conformations (looking along the C_2–C_3 bond) referred to as the *gauche* (*skew*) and *anti* (*trans*) conformations, as well as two different eclipsed conformations (Fig. 86). The two *gauche* forms are enantiomeric. The rotation about the C–C bond involves various amounts of torsional strain ('Pitzer' strain) which can be illustrated by means of a potential energy diagram (Fig. 87). The species lying at the bottom of energy minima are often referred to as 'conformational isomers' or 'conformers'. A new set of terms and symbols was suggested by W. Klyne and V. Prelog in 1960 to designate a conformational isomer for which an exact angle of torsion could not be given (Table IV).

Although the rotational barrier in ethane had been experimentally demonstrated in the 1930s, it was still unclear whether the barrier arose from repulsive or attractive forces between the hydrogen atoms; that is, whether the staggered or eclipsed form was the more stable. The quantum mechanical calculations published by Eyring in 1932 were based on the importance of both kinds of interactions but led to the prediction of a rather low potential barrier (the staggered conformation was assumed to be the more stable). In 1939, following the report of the experimental evidence that indicated a higher rotational barrier, Eyring published the results of his new theoretical calculations. This time, by considering the involvement of a resonance structure involving double bonds, the eclipsed conformation was predicted to possess the greater stability. This conclusion, however, was not in agreement with the X-ray studies of hydrocarbons, which indicated a zig-zag chain (with the hydrogens in a staggered conformation). Eyring provided a somewhat involved rationalization to account for the anomaly. The observation that cyclohexane had a lower heat of combustion than cyclopentane might lead one to expect cyclohexane to be more stable by some 3 kcal/CH_2 since all of the hydrogens are in a staggered conformation. More recently in 1958, Eyring has developed an entirely different theory. Others, such as Linus Pauling, have also attempted to provide a theoretical explanation for the existence of a rotational barrier in ethane. More recent calculations seem to provide a more satisfactory explanation of the origin of the rotational barrier; but it is difficult to give a physical picture that is satisfactory to the non-mathematician. A number of specialized reviews have been published on the topic.

B.   CYCLIC COMPOUNDS

As was seen in the previous chapter, the chemical investigations undertaken in the early part of the 20th century did not conclusively establish a multiplanar structure for cyclohexane. As was the case for the acyclic compounds, the results of the physical chemical studies on cyclic compounds were not immediately appreciated by organic chemists.

The X-ray crystallographic measurements on β-benzenehexachloride reported by Bilicke in the 1920s indicated that the carbon atoms in the ring were in a chair conformation. In 1936, K. W. F. Kohlrausch (in Graz) on examining the differences in the Raman spectra between the monosubstituted derivatives of cyclopentane and cyclohexane concluded that a substituent on cyclohexane in the chair conformation could exist in two spatial orientations (now designated as the axial and equatorial). A. Longseth and B. Bak, on the other hand, interpreted their Raman spectra a few years later in terms of a planar structure of cyclohexane. (It might be mentioned that Eyring's calculations of 1939, which predicted the greater stability of the eclipsed form of ethane, were based in part on the assumption that cyclohexane was planar.)

The measurements of the dipole moments and X-ray diffraction studies of cyclohexyl chlorides published by Odd Hassel in the 1930s provided only a limited amount of information about the structure of the cyclohexane ring. When, however, Hassel began to employ electron diffraction techniques in 1938, he soon realized that two forms of monosubstituted cyclohexane existed in a rapid equilibrium. Although the results of these studies were published in 1943 in an article entitled 'The Cyclohexane Problem' (Chapter 1), the results of Hassel's research were not widely known until after World War II when several of his articles were published in the English language. It was the conclusions published in these articles which influenced the important deductions made by D. H. R. Barton in 1950. The basic principles set forth in the 1943 paper can best be obtained by reading the paper itself, but are briefly summarized here. (The figures referred to in parentheses refer to the figures in the 1943 article.) For cyclohexane it was demonstrated that (1) the molecule existed largely in the chair form and (2) where there were substituents, there was a rapid ring inversion which allowed the substituents to occupy either the equatorial or axial position. These positions were originally designated by Hassel by the symbols 'κ' ('reclining') and 'ϵ' ('standing') [in the 1943 article: Figure 1a, b]. In 1947, Pitzer suggested the use of the terms 'polar' ('p', pointing to the pole) and 'equatorial' ('e', that is, those hydrogens lying in an equatorial belt around the carbons). Since the former term was also used to describe the electropolar nature of substituents, thus causing confusion, it was proposed in a joint communication by Barton, Hassel, Pitzer and Prelog in 1953 that it be replaced by the word 'axial'. The use of the symbols 'a' and 'e' for the axial and equatorial positions was also suggested. In 1959, the abbreviations used in the British literature were changed to '*ax*' and '*eq*'; the shorter abbreviations are still in more general use, however.

Hassel's 1943 paper also pointed out that, of the two possible conformations of monosubstituted cyclohexane, the more stable one was that in which the substituent was in the equatorial position. The reason for this was explained in 1947 with reference to cyclohexyl chloride: when the chlorine was in an axial position, there existed two hydrogens (on the 3,5-carbons) in axial positions that 'would be nearly equal to the distance between a chlorine and a hydrogen atom linked to different carbon atoms in the "*cis*" forms [that is, *eclipsed*] of chlorinated ethanes'. When the chlorine was in the equatorial position, the non-bonded distance was much greater and of the magnitude found in the staggered conformation of chloroethanes.

Even the assumed planarity of the five-membered ring began to be questioned. In 1941, J. G. Aston (at Pennsylvania State University) interpreted his entropy measurements as indicating that the cyclopentane ring was puckered (due to strong hydrogen repulsions pushing one or two carbon atoms out of the plane). A non-planar ring was also proposed by John Kilpatrick, Kenneth Pitzer and Ralph Spitzer in 1947 based on the analysis of a variety of physical data. Later (1959) Pitzer and Donath explained that although a puckered ring required an increase in angle strain, this also minimized the torsional strain. (Cyclopentane has five eclipsed ethane units of torsional strain amounting to about 14 kcal/mole.) (By the 1940s, electron diffraction studies and other physical property measurements indicated that even the four-membered ring was puckered.)

The insights of Pitzer's group in 1947 regarding the non-planarity of the cyclohexane ring seems to have been lost on most organic chemists at that time as they failed to realize the chemical significance of the conclusions. In many ways this was a bit surprising since, for example, in the photographs of the Stuart models included in the article, the steric crowding of the substituents in the cyclohexane molecules is readily apparent. These photographs might have been expected to catch the attention of organic chemists where the discussion did not.

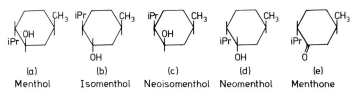

(a)	(b)	(c)	(d)	(e)
Menthol	Isomenthol	Neoisomenthol	Neomenthol	Menthone

Figure 88. Menthol diastereomers (a—d) and menthone (e).

It should be emphasized, however, that the correlation of a molecule's conformation with its chemical behavior is often a subtle one. Perhaps some organic chemists might conclude that a suggestion of a slightly puckered cyclopentane ring was a result only a physical chemist might be pleased with. The tendency on the part of most organic chemists was still to talk rather generally of the importance of steric effects rather than to look more carefully at the actual conformation of the molecule. There were, however, few experimental observations that required an interpretation involving a multiplanar cyclohexane ring — and even then alternative explanations were found. In 1934, a configurational assignment of the four diastereomers shown in Fig. 88 was made by John Read (at the University of St Andrews) based on their relative rates of reaction with p-nitrobenzoyl chloride (the relative rates were a: b: c: d = 16.5: 12.3: 3.1: 1). The formation of neomenthol (d) instead of menthol (a) on the reduction of menthone was thought to be because the 'addition of hydrogen to the keto group would be expected to take place by preference in the region removed by the large isopropyl group'. It was not so easy to explain the relative rates of all four compounds with p-nitrobenzoyl chloride, since it might appear that the hydroxyl group in the isomenthol compound (b) was the least sterically hindered. This problem was apparently not appreciated by Read.

The solution to these kinds of stereochemical problems was made possible much later after the publication of Barton's paper in 1950. To

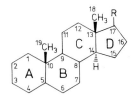

Figure 89. Steroid nucleus.

understand the significance of this paper (reprinted in Chapter 1), some additional background is needed:

In the late 1940s, a large number of organic chemists were involved in the synthesis and determination of the structure of a variety of steroids. The steroid nucleus is a polycyclic compound containing three six-membered (A, B, C-rings) and one five-membered ring (D-ring) (Fig. 89). The steroid structures are usually illustrated with all rings coplanar, with the methyl groups at carbons 10 and 13 above the plane of the rings. In 1930, Louis F. Fieser (at Harvard University) designated those substituents situated above the plane as 'β'-oriented; they were represented by a solid, heavy line. Substituents below the plane were 'α'-oriented and the bond was indicated with a dotted line. Although this method of representation was quite satisfactory in picturing the relative configuration of stereoisomers, the picture could not account for a number of stereochemical observations. For example, in the reaction of acetyl chloride with cholic acid, a steroid derivative having α-oriented hydroxyls on C_3, C_7 and C_{12} (Fig. 90), it was observed that the C_3 and C_7 hydroxyls were esterified whereas the C_{12} hydroxyl was untouched. It was difficult to understand the reasons for this difference in reactivity since all three hydroxyls were oriented beneath the plane of the rings and apparently situated in a similar steric environment. Such observations were only explained by Fieser rather broadly in terms of the operation of some sort of steric effect. In 1950, for example, in the paper that precedes Barton's paper in *Experientia*, Fieser could do no more than classify the types of steric hindrance that might be important. He used the term 'intraradial' or 'interradial' effects to indicate whether certain substituents seemed to

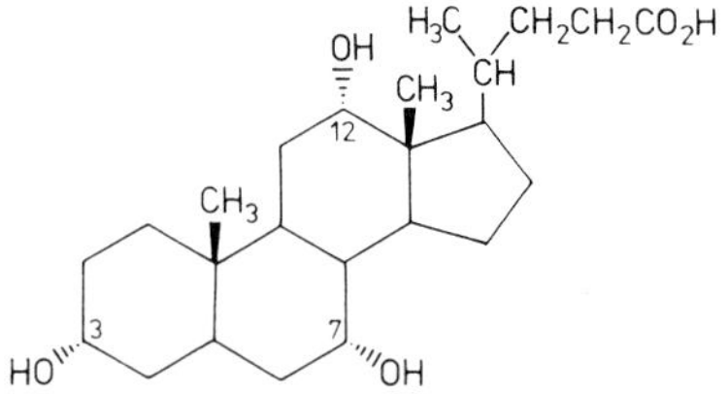

Figure 90. Cholic acid.

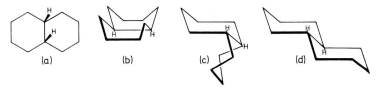

Figure 91. Conformations of decalin. *cis*-Decalin: (a) planar, (b) 'double-boat' or 'cradle', (c) 'double-chair'. *trans*-Decalin: (d) 'double-chair'.

produce the steric effects when either within the Van der Waals radius of the reacting center, or outside of it.

Derek Barton was able to provide a solution to these kinds of problems because of his background of research in both steroid chemistry and physical chemistry. In 1945 he had been studying molecular rotation differences in steroids and as a result of this work he became interested in a paper published by O. Bastiansen and O. Hassel in 1946. Their electron diffraction studies indicated that *cis*-decalin had a 'double-chair' structure (Fig. 91c) rather than the 'double-boat' structure suggested by Mohr sometime earlier (Figs. 79, 91b). Barton had also studied the papers published at the same time by Dostrovsky, Hughes and Ingold, and by Westheimer and Mayer, who had attempted a semiquantitative method of calculating the steric effects in bimolecular substitution reactions due to non-bonded interactions. This method was used by Barton in 1948 to calculate the preferred conformations of ethane, cyclohexane and the *cis*- and *trans*-decalins. A set of accurate models was designed for this purpose (an approach taken by Hermans some 25 years earlier). Table V summarizes the results of Barton's calculations.

TABLE V
Relative conformational stabilities (Figure illustrations provided in parentheses)

Compound	Relative stability
Ethane	staggered > eclipsed (85)
Cyclohexane	chair > boat (95)
Decalin	*cis* (91c) > *cis* (91b)
	trans (91d) > *cis* (91b, c)

In the same year Barton used conformational arguments to explain the magnitude of the ionization constants of the tricarboxylic acid formed from the oxidation of abietic acid, and thereby decided on the stereochemistry of the ring fusion in abietic acid. Two years later he demonstrated how a knowledge of the conformation of the steroid nucleus could be used to understand the stereochemistry of an elimination reaction. When cholesterol is brominated, there is formed what was called the ordinary 'labile' dibromide ($5\alpha,6\beta$), which on standing gives a 'stable' dibromide ($5\beta,6\alpha$). The ordinary dibromide reacted rapidly with potassium iodide to regenerate cholesterol, while the stable dibromide did not react under the same conditions. The reason for the difference in reactivity was only apparent when it was recognized that the rings were not coplanar, but were in chair conformations. Sometime earlier, Ingold and Hughes had shown that in bimolecular eliminations (E_2) the groups to be eliminated must be coplanar (termed an *anti* elimination). In the 'labile dibromide', the two bromines are found in a diaxial conformation favorable for the *anti* elimination; in the 'stable' dibromide, the two bromines are in equatorial positions, a conformation unfavorable for an *anti* elimination. The stereochemistry of elimination reactions in cyclic compounds is still an interesting area of research but will not be discussed in further detail here. (It might be noted that the mechanism of the base-catalyzed elimination reaction of β-benzenehexachloride, in which all of the chlorines are in equatorial positions, was discussed with reference to a planar cyclohexane ring well into the late 1950s.)

While Barton was a visiting lecturer at Harvard University in 1949–50, he attended a lecture in which Fieser discussed the difficulty he had in understanding the differences in rates of esterification and oxidation of different hydroxyl groups in steroids. Barton's previous research studies enabled him to see immediately that the answer must be found by considering the conformation of the steroid rings. In the case of cholic acid, discussed earlier, he could demonstrate that the three hydroxyl groups were in different steric environments. The 3α-hydroxyl was in an equatorial position whereas the 7α- and 12α-hydroxyls were both in axial positions and thus more hindered (due to 1,3-non-bonding interactions) than the 3α-hydroxyl. The differences in reactivity of the 7α- and 12α-hydroxyl groups, observed in other reactions, were attributed to the differing number and kind of these 1,3 interactions. In Barton's 1950 paper are found numerous examples of this sort which demonstrated how a knowledge of the conformation of a molecule could be used to explain the differences in the stability or reactivity of stereoisomers.

The publication of this paper, as well as the numerous lectures given by Barton in the USA in that year, had a dramatic effect upon the subsequent development of stereochemistry. In a 1951 review of the 'Conformation of Cyclic Systems' appearing in the *Annual Reports of the Chemical*

Society, A. J. Birch stated: 'Conformational analysis for the study of the stability and reactivity of saturated or partly saturated cyclic systems promises to have the same degree of importance as the use of resonance in aromatic systems.' By 1957, the *Experientia* paper had been designated by one reviewer as a 'classic'. In 1959, another wrote: 'so many applications of conformational analysis at present appear in the literature that the method by now may be regarded as a thoroughly integrated part of theoretical organic chemistry'. [However, the impact of the articles was not immediately felt in all areas of chemistry. For example, none of the speakers (including Pitzer) mentioned Barton's paper at a Faraday Society meeting (at Oxford in April 1951) concerned with 'Hydrocarbon Structure and Bond Properties'. Even Hassel's ideas were only briefly mentioned in one of the general discussions.]

It is difficult to understand why it was that Barton's paper had such a great impact, since in the previous decade other papers had been published that also stressed the importance of considering the relationship between molecular conformation and chemical reactivity. Prelog, for example, had published several papers concerned with the conformation of medium-sized rings. Richard Reeves (at the Southern Regional Research Laboratory of the US Department of Agriculture) had examined the relationship between the conformation of glycosides and the ease of formation of cupraammonium complexes. These studies seem to have been well known, at least to specialists in the fields, but the general approach to conformational analysis does not seem to have been appreciated. Although Barton's paper was concerned with the conformational analysis of steroids, perhaps the general applicability of his approach was appreciated because of his emphasis on the difference in the reactivity of groups in the equatorial and axial positions.

Barton was able to illustrate the greater reactivity of substituents in the equatorial position relative to the axial by examining the stereochemistry of reactions of rigid bi- and polycyclic systems in which there was no

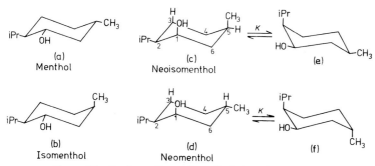

Figure 92. Conformations of menthol stereoisomers.

mobility of the substituents, that is, in compounds in which there could be no change from the equatorial to the axial due to chair–chair inter-conversions. In 1953, Ernest Eliel (at the University of Notre Dame) re-examined the earlier observations of John Read (Fig. 88), concerning the relative reactivities of isomeric menthols, and raised the question of how one would be able to decide whether the greater reactivity of a reactant in the equatorial position would still be preserved in readily inverted ring systems. If the isomeric compounds react through their most stable con-formations (*a, b, c, d* in Fig. 92), it is not difficult to understand why *a* and *b* react more rapidly than *c* and *d*, since the hydroxyl groups in the former compounds are in equatorial positions. But on this basis we would also predict that *d* should be more reactive than *c* since in the latter com-pound there is a non-bonded crowding of a methyl group and hydrogen of C_3 and C_5, but only hydrogens at these positions in *d*. The observation that the relative rates are in fact reversed might be explained by assuming that the reaction occurred by way of a conformation in which the hydroxyl group was in an equational position (*e* and *f*) and the difference in rates reflects the lower equilibrium concentration of *f* compared to *e* (since to produce conformation *f*, both alkyl groups must be placed in axial positions).

In this system there was no way of knowing for sure whether in fact this was the case. In 1954–55, a general equation was derived independ-ently by S. Winstein and N. Holness, and E. L. Eliel and C. A. Lukach, to describe these types of mobile systems. For a monosubstituted derivative of cyclohexane, in which a reactive group, X (Fig. 93), could be in an axial or equatorial position, the specific rate constant (*k*) can be expressed in terms of the rate constants of X in the equatorial (k_e) and axial (k_a) positions, and the equilibrium constant (*K*) or mole fraction (*N*) of each. Winstein and Holness suggested an experimental method for determining the value of k_a and k_e. In cyclohexyl derivatives containing a 4-*tert*-butyl substituent, the bulk of the alkyl group was so large that it was assumed that it must be in an equatorial position. As an example, in *cis*- and

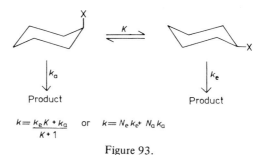

$$k = \frac{k_e K + k_a}{K + 1} \quad \text{or} \quad k = N_e k_e + N_a k_a$$

Figure 93.

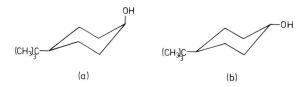

Figure 94.   The most stable conformations of (a) *cis*- and (b) *trans*-4-*t*-
butylcyclohexanol.

trans-4-*tert*-butylcyclohexanol (Fig. 94) the hydroxyl groups would exist, respectively, only in the axial and equatorial positions. The rate constants obtained for the reactions (such as in esterification) of the compounds could be used to assign the values of k_a and k_e in unsubstituted cyclohexanol. Although the use of *tert*-butyl groups to lock rings into a known conformation was later shown to oversimplify the conformation problem (because the implicit assumption that the *t*-butyl group does not affect the specific rate of the equatorial and axial conformers is not accurate), the approach did serve as an important stimulus to the subsequent development of conformational analysis. A rather interesting observation in this regard is that reported in 1960 by N. L. Allinger and L. A. Freiberg, who demonstrated that *trans*-1,3-di-*tert*-butylcyclohexane could not exist in a chair conformation since this produced an intolerable interaction between one of the *tert*-butyl groups in an axial position with other axial hydrogens. The compound, therefore, is found in a boat conformation.

Recent conformational studies have indicated that cyclohexane cannot be considered only in terms of the chair and boat conformation. 'Twisted' conformations are also observed. The existence of these forms had been anticipated by Sachse and Mohr, and later discussed by Wightman, Hückel, Brodetzky and Henriquez in the 1920s. W. A. Wightman (at the University of Leeds) demonstrated with molecular models how easily the flexible boat form could be converted to other strain-free forms by rotations about carbon–carbon single bonds. Most organic chemists, however, considered only the chair and boat forms (as well as the planar structure) because these were the forms easily visualized with models; others appear only as transitional forms and therefore might be thought incapable of a separate existence. By the 1960s, the conformation of a molecule was described not in terms of the predictions based upon a mechanical model, but in terms of the total internal energy of the molecule, which (on top of the Pauling bond energy) was the resultant of: (1) the bond stretching and compression energy, (2) the angle deformation energy (Baeyer strain), (3) torsional energy (Pitzer strain), (4) non-bonded energy (Van der Waals attraction and repulsion), (5) dipolar interaction energy, including that due to intramolecular hydrogen bonds, and (6) solvation energy. The

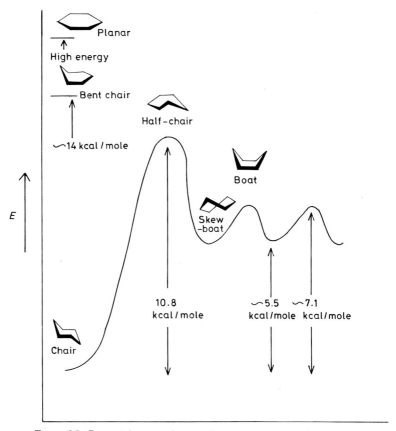

Figure 95. Potential energy diagram for cyclohexane conformations.

molecule adjusts its conformation in such a manner that the sum total of these energies is at a minimum. In 1951, P. Hazebroek and L. J. Ooster-hoff thus calculated that the chair form was more stable than the boat form by about 5.5 kcal/mole. These calculations also revealed that the boat form was only one of several possible flexible conformations. Although the boat form is free from angle strain, it still contained con-siderable bond eclipsing (four pairs of hydrogens on the 'sides' of the boat) as well as a Van der Waals interaction between the two hydrogens occupying the 'bowsprit' or 'flagpole' positions (see Appendix C). By rotating about C–C single bonds, so as to pass from one boat to another, a conformation is obtained in which these two kinds of interactions are minimized. This conformation is called the 'skew-boat' or 'twist form' (sometimes the 'flexible form'), and represents the true energy (more stable than the boat by some 1.6 kcal/mole) of the flexible form. The computer calculations reported by James Hendrickson (at UCLA) in 1961 provided an even more detailed description of the conformations of cyclohexane. These calculations indicated that the boat form represented a *maximum* in a potential energy diagram (Fig. 95) with the skew-boat occupying a position of lower potential energy. The transition from the chair to the skew-boat passes through an energy maximum in which the molecule is in what has been termed the 'half-chair'. The half-chair form has four atoms of the ring coplanar and the other two above and below the plane of the other four carbons. The 'bent-chair' form, which is produced by bending up one carbon of the chair into a position coplanar with four other carbon atoms, would be of an even higher energy than the half-chair and therefore is probably not involved in any interconversions of the cyclohexane ring. The planar cyclohexane ring has an even higher energy content, since it contains not only torsional strain (since all of the hydrogens are eclipsed) but considerable angle strain. It is somewhat ironic to note that two of the conformations of cyclohexane that have played such a central role in the earlier history of conformational analysis have now been relegated to either a transient existence (the boat form) or are not considered to be involved at all (the planar form) in the interconversions of cyclohexane conformations.

15

Asymmetric transformations

Much of the history of stereochemistry can be traced back to the early investigations of optical activity and optical isomerism. Quite early in the history of these investigations it was also noted that in Nature only one of the optical isomers was found, usually to the exclusion of the other. The theories and experimental investigations that attempted to explain the formation or destruction of one isomer in a living or non-living system are the subject of this chapter. In contemporary accounts of stereochemistry, asymmetric transformations are often discussed in terms of three categories: asymmetric synthesis, asymmetric destruction, and kinetic methods of resolution. (The definitions of these terms are given in the Glossary, Appendix B.)

In 1858, Pasteur had noticed that when a solution containing the ammonium salt of (±)-tartaric acid was allowed to ferment in the presence of a mould of *Penicillum glaucum*, the (+)-tartaric acid was used up, leaving the (−)-tartaric acid in the fermentation broth. This is an example of what is now termed an asymmetric destruction. Since Pasteur did not believe that the cause of the destruction of one enantiomer could be related simply to the asymmetry of the tartaric acids themselves, he suggested that a special force was responsible for the destruction of one enantiomer:

> Thus we find introduced into physiological principles and investigations the idea of the influence of the molecular asymmetry of natural organic products, of this great character which establishes perhaps the only well-marked line of demarcation that can at present be drawn between the chemistry of dead matter and the chemistry of living matter.

Pasteur's revival of a vitalistic view was apparently supported by the fact that no optically active, or even racemic, compound had been synthesized in the laboratory in the absence of a living system. The experimental basis for Pasteur's view was seriously undermined in 1873 with E. Jungfleisch's report of the preparation of racemic tartaric acid from 'synthetic' succinic acid. This resulted in only a slight modification in Pasteur's ideas:

> . . . this barrier still exists. . . . To transform one inactive compound into another inactive compound which has the power of resolving itself simultaneously into a right-hand compound and its opposite, is in no way comparable with the possibility of transforming an inactive compound into a single active compound. This is what no one has ever done; it is, on the other hand, what living nature is doing increasingly before our eyes.

Within another decade, however, Pasteur finally admitted that there was no absolute barrier between the living and non-living systems.

The concept of asymmetric synthesis was first enunciated in 1874 by Le Bel:

> When an asymmetric body is formed in a reaction where there are present originally only symmetrical bodies, the two isomers of inverse symmetry will be formed in equal quantities . . . this is not necessarily true of asymmetric bodies formed in the presence of other active bodies, or traversed by circularly polarized light, or, in short, when submitted to any cause whatever which favors the formation of one of the asymmetric isomers. Such conditions are exceptional; and generally in the case of bodies prepared synthetically those which are active will escape the observation of the chemist unless he endeavors to separate the mixed isomeric products, the combined action of which upon polarized light is neutral.
>
> We have a striking example of this in tartaric acid, for neither the dextro- nor the levo-tartaric acid has ever been obtained directly by synthesis, but the inactive racemic acid which is a combination of equal parts of the dextro- and levo-acids, is always obtained.

Le Bel's perceptive comments do not seem to have stimulated any experimental investigations or controversy. The first concrete suggestion of how an optically active substance (such as the sugars) might be formed in living systems was proposed by Emil Fischer in 1890:

> No fact hitherto known speaks out against the view that the plant, like chemical synthesis, first prepares the inactive [race-

mic] sugars; that it then resolves them into their active constituents.

According to Fischer, more complex molecules were then built up from
the simpler optically active substances obtained in this 'resolution' step.
The presence of racemic lactic acid in sour milk was cited as an example
of such an 'unresolved' compound. The selective consumption of one
enantiomer was viewed by Fischer as requiring the stereochemical action
of an enzyme:

> The enzyme and glucoside must fit together like a key and lock,
> in order that one may exercise a chemical action on the other.

Fischer proposed the 'key and lock' theory to account for the observation
that only the enzyme *maltase*, but not *emulsin*, catalyzed the hydrolysis
of α-methylglucose; the reverse was the case for the hydrolysis of β-
methylglucose. Fischer recognized that not all optically active substances
were produced by a destructive process; they might be produced by a
'selective production' under asymmetric influences. Fischer pointed to the
cyanohydrin synthesis of sugar stereoisomers as an example of a selective
production.

In 1899, Willy Marckwald and Alexander McKenzie in Berlin developed
a new method of resolving organic compounds. The method was described
as an example of asymmetric synthesis, although it would now be classified as an example of a kinetic method of resolution. They discovered that
when ($\pm$)-mandelic acid was heated with ($-$)-menthol, the ($-$)-mandelic
acid was not esterified as rapidly as the (+)-acid. In 1904, Marckwald proposed that the term 'asymmetric synthesis' be used to refer to the preparation of an optically active compound from a symmetrical substance via an
optically active intermediate. An example of such a synthesis was furnished by Marckwald when he observed that a slightly levorotatory sample of
2-methylbutanoic acid (containing a new asymmetric carbon) was produced when the brucine salt of methylethylmalonic acid was decarboxylated:

$$[CH_3CH_2(CH_3)C(CO_2H)(CO_2)]^{-1} \ [C_{23}H_{26}N_2O_4H]^{+1} \xrightarrow{\ -CO_2\ }$$
$$\overset{*}{CH_3}CH_2(CH_3)CHCO_2H$$

When McKenzie returned to the University of Birmingham, he continued
the studies he had initiated with Marckwald earlier. Examples of the asymmetric syntheses (now referred to as 'asymmetric inductions') studied

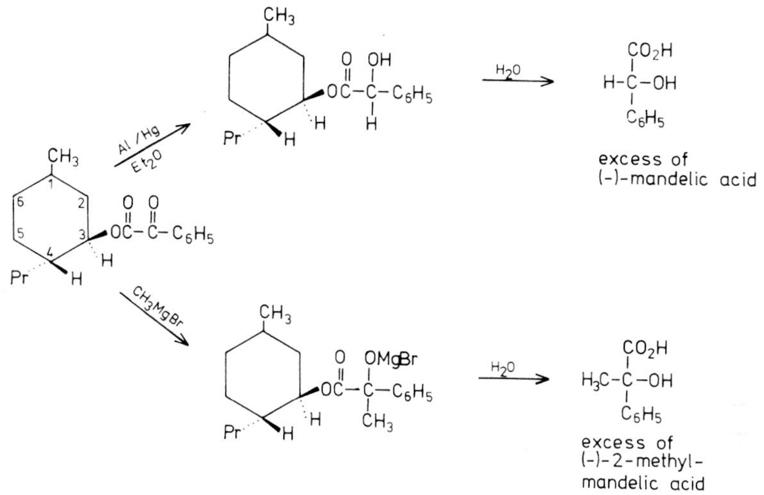

Figure 96.

by McKenzie in 1904 are shown in Fig. 96. The reactions showed how the stereochemistry involved in the creation of a new asymmetric center may be influenced by the presence of another asymmetric center. In these examples, the (−)-menthylphenylglyoxylate contains two asymmetric carbons (C_3 and C_4). Although it was clear that the excess of one enantiomer of mandelic acid or 2-methylmandelic acid was controlled by the asymmetry of the starting material, it was not possible to predict which enantiomer would be preferentially formed. It was not until the 1950s that V. Prelog and D. J. Cram independently showed how such a prediction could be made providing information was available about the absolute configuration of the reactant. In the compounds studied by Prelog, the original asymmetric carbon is usually found to be somewhat distant from the site at which the new asymmetric carbon is generated. To generalize, the most stable conformation of a compound having the formula R–C–C–O–C(S)(M)(L) will be determined by the size of the three groups attached to the asymmetric carbon atom. (The relative size of the groups is indicated by the letters S, M, L meaning 'small', 'medium' and 'large', respectively.) The reagent can be expected to approach the

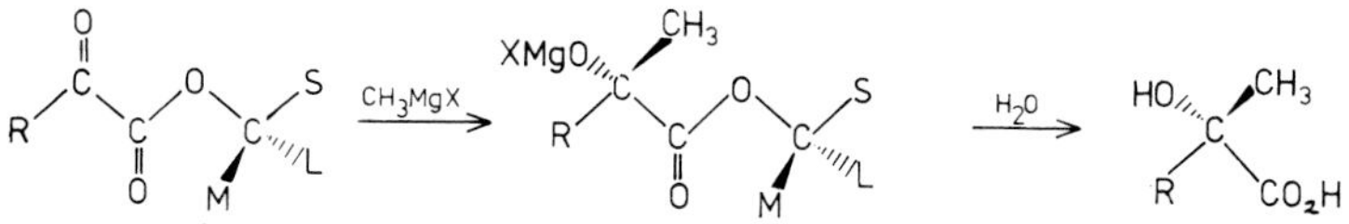

Figure 97. Illustration of the application of 'Prelog's rule'.

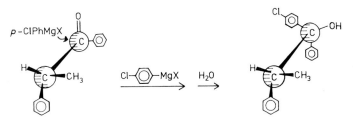

Figure 98. Illustration of the application of 'Cram's rule'.

ketone carbonyl group from the side containing the smaller of the remaining two groups (Fig. 97). The reaction of (−)-menthylphenylglyoxylate with methyl magnesium bromide (Fig. 96) conforms to Prelog's rule if you consider that S = H (on C_3), M = the methylene group (C_2), and L = the groups about C_4 in the menthol ring.

Cram's rule is concerned with the stereochemistry of reactions in which the original asymmetric carbon and the new one are found adjacent to each other:

> In non-catalytic reactions of the type shown below [Fig. 98] that diastereomer will predominate which would be formed by the approach of the entering group from the least hindered side of the double bond when the rotational conformation of the C–C bond is such that the double bond is flanked by the two least bulky groups attached to the adjacent asymmetric center.

The question of the ultimate origin of optical activity has been of interest to scientists for some time. In 1888, Pasteur considered that dissymmetric (chiral) substances produced in Nature could be attributed to two causes: either the synthesis took place in the presence of some other chiral substance, or the synthesis involved a chiral physical force. Any synthesis involving chiral forces is generally referred to as an 'absolute asymmetric synthesis'. Pasteur himself spent a number of years attempting such syntheses − for example, by running reactions in rotating tubes or studying the effect of a strong magnetic field on the crystallization of racemic mixtures. (It was later shown by Pierre Curie that the external influences employed by Pasteur were not chiral.)

In 1898, F. R. Japp (who was then president of the Chemical Section of the British Association for the Advancement of Science) reintroduced vitalistic arguments to explain the original syntheses of optically active compounds:

> The absolute origin of compounds of one-sided symmetry found in the living world is a mystery as profound as the origin

of life itself . . . the production of a single enantiomorph cannot conceivably occur through the chance play of symmetric forces . . . I see no escape from the conclusion that, at the moment when life first arose, a directive force came into play — a force of precisely the same character as that which enables the intelligent operator, by the exercise of his Will, to select one crystallized enantiomorph and reject its asymmetric opposite.

The controversy elicited by Japp's remarks cannot be discussed here. Suffice it to say, it was not long before the results of a number of experimental studies had been published that demonstrated the *in vitro* synthesis of optically active substances.

The spontaneous generation of an optically active substance from a racemic mixture is not as mysterious as the synthesis of the substance in a living system. Kipping and Pope had shown in 1898 that when a saturated solution of sodium chlorate was allowed to crystallize spontaneously, only one or two of the forty-six trials produced an equal number of dextrorotatory and levorotatory crystals. Although the average number of each kind of crystals in all of the trials was close to 50%, the amount of the dextrorotatory crystals varied from 24% to 77% in individual trials. When the crystallization was allowed to take place in a 20% (+)-glucose solution, there were twice the number of levorotatory crystals produced as the dextrorotatory ones. The glucose in the solution was therefore functioning as an asymmetric influence on the crystallization. A glucose solution also had a similar effect upon the crystallization of sodium or ammonium tartrate from a racemic mixture.

An interesting variation of these studies was reported by E. Havinga at Leiden in 1941. When allylethylmethylanilinium iodide was allowed to crystallize from a chloroform solution, the crystals obtained were usually optically active — sometimes dextrorotatory, sometimes levorotatory:

$$[CH_3-\overset{\overset{\displaystyle C_2H_5}{|}}{\underset{\underset{\displaystyle C_6H_5}{|}}{N}}-CH_2-CH=CH_2]^+ \ I^- \ \rightleftharpoons \ CH_3-\overset{\overset{\displaystyle C_2H_5}{/}}{\underset{\underset{\displaystyle C_6H_5}{\backslash}}{N}} \ \rightleftharpoons \ [CH_2=CH-CH_2-\overset{\overset{\displaystyle C_2H_5}{|}}{\underset{\underset{\displaystyle C_6H_5}{|}}{N}}-CH_3]^+ \ I^-$$

$$+$$

$$CH_2=CH-CH_2I$$

dextro–crystals   *levo*–crystals

The interconversion of enantiomers was possible because of the equilibrium involving a dissociation to form the amine and allyl iodide. The fact that either enantiomer could be produced suggested that dissymmetric contaminants did not promote a preferential nucleation or growth

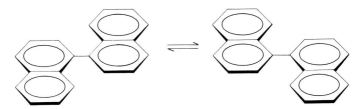

Figure 99.   Binaphthyl equilibrium: crystallization of the conformer on
the left gives (−)-binaphthyl; that on the right, (+)-binaphthyl.

of one enantiomer, but this possibility could not be excluded. A similar, but more statistically persuasive experiment was demonstrated by H. Pinnock in 1971 in his study of the crystallization of binaphthyl from a melt at 150 °C (Fig. 99). It was found that from two hundred separate runs, the specific rotation of the sample obtained varied from −218° to +206° with the mean, however, being about +0.14°. (At room temperature, solutions of binaphthyl are optically stable.)

In 1894, Van't Hoff proposed that optically active substances might be synthesized in the presence of circularly polarized light. The early experiments that were run to test this idea were unsuccessful, because in the first reactions studied, light was not essential for the reaction. It was not until 1929–30 that Werner Kuhn and his students were able to demonstrate such a synthesis. (The reactions studied were, however, examples of 'asymmetric destructions' rather than 'asymmetric syntheses'.) When the ethyl ester of 2-bromopropanoic acid, $CH_3CH(Br)CO_2CH_2CH_3$, or the dimethylamide of 2-azidopropanoic acid, $CH_3CH(N_3)CON(CH_3)_2$, were partially decomposed by circularly polarized light, the undecomposed compounds were found to exhibit a small amount of optical activity. In 1935, Tenney Davis (at MIT) reported the isolation of an optically active product from the (circularly polarized) light-catalyzed bromination of 2,4,6-trinitrostilbene. He concluded that this showed that 'the ethylenic linkage of trinitrostilbene is by no means the rigid structure conceived by Van't Hoff and suggests that it is in some manner potentially asymmetric'. Ten years later, Davis reported the formation of diethyl-(+)-tartrate when diethyl fumarate was hydroxylated in a reaction catalyzed by dextrorotatory circularly polarized light. It has been suggested that since the sunlight reflected by the sea possesses a slightly circular (or elliptical) polarization, asymmetric syntheses could have taken place on the surface of the earth millions of years ago.

Those interested in the question of the origin of asymmetry in Nature are usually not so much concerned with how the asymmetry originated (which could happen by a chance asymmetric crystallization, for example) but in understanding why only one type of chirality (the L-amino acids, the D-sugars) seems to predominate. If the reflected light is always circularly polarized in the same direction then perhaps we have found an explanation. An alternative view is that both types of chiral molecules were produced originally but that in the evolutionary scheme one kind won out over the other. This is suggested by the presence of D-amino acids in some lower forms of life such as bacteria and antibiotics.

Other forms of asymmetric physical forces might also be involved in absolute asymmetric syntheses or destructions. For example, in 1968 A. S. Jaray in Hungary reported that the polarized γ-rays produced from radioactive sources resulted in the preferential destruction of the D-isomer in racemic solutions of tyrosine.

Only a few examples of the earlier work relating to relative and absolute asymmetric syntheses have been mentioned in this chapter. The reader is urged to consult some of the more specialized publications in order to appreciate the importance of the numerous studies that have appeared in recent years.

16

Some recent developments and future expectations

Since the 1950s the incorporation of stereochemical concepts into all areas of chemical research has been so rapid and widespread that it is difficult to summarize the present state of stereochemistry, much less the future. The discussion that follows is, therefore, far from comprehensive, but it is hoped that the areas of stereochemical research that are discussed are those that will continue to develop in the future. Since this discussion generally focuses on particular examples taken from a broad area of stereochemical research, the historical development may be incomplete.

A. 'CHEMICAL CURIOSITIES'

In the past, as in the present, many organic chemists have attempted the synthesis of molecules having unusual shapes or properties. Generally it was hoped that the preparation of these molecules would either confirm, extend or challenge the existing tenets of stereochemistry. The preparation of optically active allenes, compounds containing asymmetric hetero atoms, and biphenyl derivatives discussed in earlier chapters might be cited as examples of research of this kind. Although by the mid-20th century many chemists had turned their attention to the more dynamical aspects of stereochemistry, a large number of publications continued to appear in which there were reported the syntheses of molecules having some unlikely twisted and strained structures. Although some chemists feel that few stereochemical insights will be gained by the synthesis of such molecules, others (such as the Crams in California) argue that the reports are a sign of the maturation of organic chemistry:

> As the structural theory of organic chemistry has reached maturity, more investigations have been directed toward defining its boundaries. Experimentalists have synthesized internally

tortured molecules with inherent suicidal tendencies that skirt a fine line between isolability and self-destruction. Theorists have progressed from rationalizing what is observed to predicting what extremes might be incorporated into organic structures before molecular collapse or fragmentation occur.

It seems almost inconceivable that the bond angles of the bridging carbon atoms in the cyclophanes (Fig. 100) could be so distorted to allow the two aromatic rings to lie in such close proximity. The *ortho-, meta-* or *para*-carbons can be joined with carbon–carbon bonds or double bonds, or even multiple bridges. One of the goals of this area of research is to synthesize the fully bridged unsaturated cyclophane (Fig. 100b). In 1979, Virgil Boekelheide reported the preparation of the saturated cyclophane. Achiral and chiral multilayered paracyclophanes have also been synthesized (Fig. 101a and b, respectively).

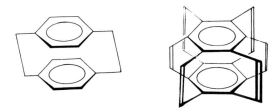

Figure 100.   Some cyclophanes: (a) a paracyclophane, (b) a completely bridged, unsaturated cyclophane.

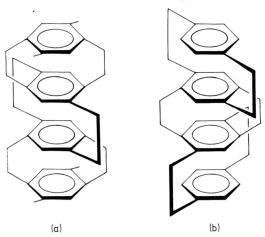

(a)   (b)

Figure 101.

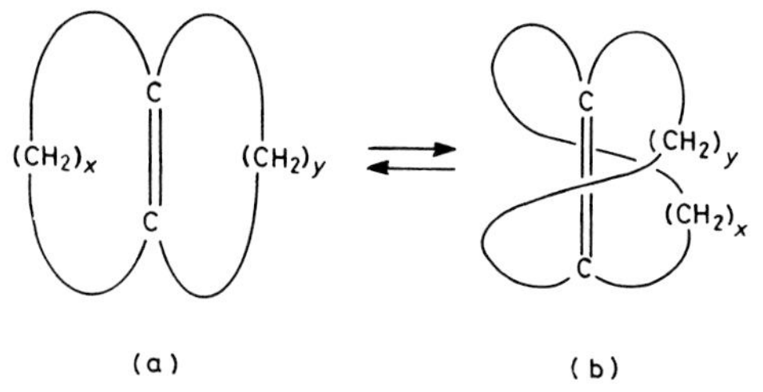

(a)   (b)

Ring size	Compound (a)	Compound (b)
$x = 8, y = 10$	cis−bicyclo[10.8.0]eicos-1(12)−ene	[10.8] betweenanene
$x = 10, y = 10$	cis−bicyclo-[10.10.0]docos−1(12)−ene	[10.10] betweenanene

Figure 102.

The nature of the interaction of the π-system is obviously affected by the spatial orientation and proximity of the aromatic rings. The chemical reactivity of functional groups can be markedly affected by its structural environment. Accordingly, a number of molecular curiosities have been prepared which exhibit unusual chemical behavior. In 1977, for example, James Marshall (at Northwestern University) synthesized a compound which he described as a 'betweenanene' (Fig. 102b). The name was derived from the spatial arrangement of the lower members of the family wherein the double bond ('ene') is sandwiched 'between' the alkyl ('ane') chains. It is not possible to name the compound using the E, Z convention, if the two chains are identical ($x = y$). The 'cis' compound (102a) isomerizes to the more stable betweenanene in the presence of acid (70% 102b for $x = 8$, $y = 10$; 95% 102b for $x = y = 10$). The isomerization is believed to involve the protonation of the double bond, indicating that a 'small' reactant can penetrate the surrounding cage. The betweenanenes are quite stable to 'larger' reactants such as m-chloroperoxybenzoic acid, which converts 102a to an oxide with ease.

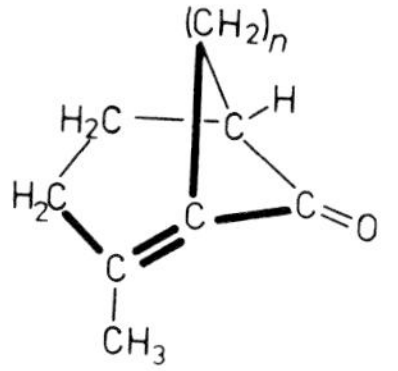

Figure 103.

In 1913, J. Bredt, after studying the reactions of a number of bicyclic compounds, formulated a rule which stated that a double bond could not be formed at a bridgehead carbon atom. In 1946, V. Prelog was able to show that the rule did not apply if the bridgehead carbons were part of a sufficiently large ($n \geqslant 8$) ring system (Fig. 103). (See also structure II in

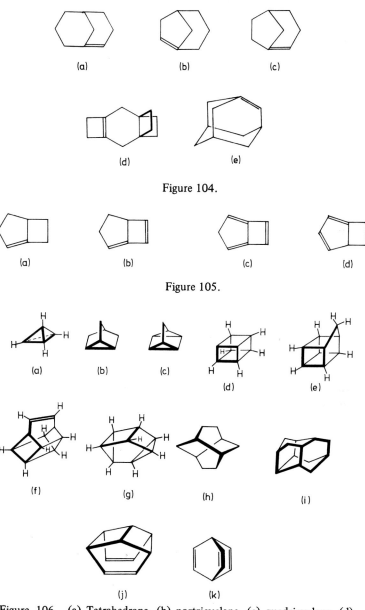

Figure 104.

Figure 105.

Figure 106. (a) Tetrahedrane, (b) nortricyclene, (c) quadricyclene, (d) cubane, (e) homocubane, (f) basketene, (g) cuneane, (h) twistane, (i) congressane (the symbol of the IUPAC Congress), (j) hypostrophene, (k) barrelene.

the 'Bredt's rule' figure in Prelog's article.) The synthesis of stable 'Bredt compounds' containing only seven bridging carbons (Fig. 104 a–c) in 1967 was thought to be the lower limit until the synthesis of compounds 104d and e. Reports continue to appear on the synthesis of small, unsaturated bicyclic compounds (Fig. 105) whose stability seems to defy what would appear to be their 'strained' character. The past two decades have seen the publication of reports of the syntheses of compounds which are best described as 'chemical curiosities' (Fig. 106). The 'Platonic molecules' (see Fig. 1) tetrahedrane (Fig. 106a), cubane (Fig. 106d) and dodecahedrane have been prepared (respectively by, G. Maier and S. Pfriem, 1978; P. Eaton and T.W. Cole Jr, 1964; L. Paquette and co-workers, 1981). Dodecahedrane, $C_{20}H_{20}$, is a hydrocarbon-polyhedron constructed with 12 cyclopentane rings.

In hypostrophene (j) and barrelene (k), the nature of π-bond interaction across space could be studied. Although barrelene is isoelectronic with benzene, it shows no aromatic character. Derivatives of the valence-bond isomers of benzene: 'Dewar' benzene, prismane and benzvalene (Fig. 107) have been prepared but the unsubstituted compounds are unstable. In 1971, Thomas J. Katz, E. J. Wang and N. Cloton at Columbia University reported concerning the benzvalene synthesis: 'However, we were discouraged from investigating its large-scale isolation further when the resonance energy of benzene [to which it isomerizes] and the strain energy of small-ring compounds were called to our attention by the benzvalene exploding.' Perhaps they also were discouraged by the 'extraordinary foul odor' of benzvalene. Electron diffraction studies of the

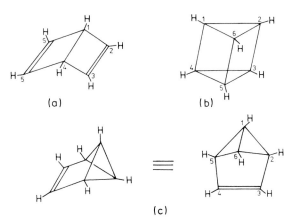

Figure 107. (a) 'Dewar' benzene (bicyclo [2.2.0] hexa-2,5-diene), (b) prismane (tetracyclo $[2.2.0.0.^{2,6}0^{3,5}]$ hexane, (c) benzvalene (tricyclo $[3.1.0.0^{2,6}]$ hex-3-ene).

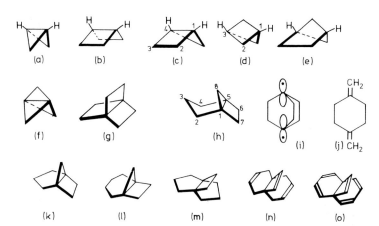

Figure 108. (a) Bicyclo [1.1.0] butane, (b) bicyclo [2.2.0] hexane, (c) bicyclo [2.1.0] pentane, (d) bicyclo [1.1.1] pentane, (e) bicyclo [2.1.1] hexane, (f) tricyclo $[1.1.0^{3,5}]$ pentane or [1.1.1] propellane, (g) tricyclo $[2.2.0^{1,4}]$ octane or [2.2.2] propellane, (h) tricyclo $[3.2.1.0^{1,5}]$ octane or [1.2.3] propellane; (i)–(j) other intermediates or products in propellane synthesis; (k)–(o), some other saturated and unsaturated propellanes.

stable hexamethyl derivative of 'Dewar' benzene have shown that the central bond is one of the longest carbon–carbon single bonds (1.63 Å) known.

Another class of compounds that have been the object of extensive experimental and theoretical studies are the small bicyclo and tricyclo hydrocarbons (Fig. 108). The tricyclic compounds with a zero bridging are commonly referred to as 'propellanes'. The bridging single bond ($C_1 - C_4$) in bicyclo [2.1.0] pentane of 1.44 Å is reported to be the shortest carbon–carbon single bond known, while the $C_2 - C_3$ bond of 1.62 Å is almost the longest. On the other hand, the $C_1 - C_3$ distance of 1.84 Å in bicyclo [1.1.1] pentane (108d) is considered to be the smallest 'non-bonded' distance between carbons. *Ab initio* molecular orbital theory indicates that the bridgehead carbon atoms in the bicyclo hydrocarbons are essentially sp^2-hybridized, and that the central bond is due to the overlap of p-orbitals. The propellanes could be considered to be derived from the overlap of p-orbitals in a diradical. However, in reactions that produced the diradical (108i) the product was dimethylenecyclohexane rather than the propellane (108g). The smallest stable propellane that has been synthesized thus far is the [1.2.3] propellane (108h). Although [1.2.3] propellane (108h) is achiral (it possesses a plane of symmetry), it has been reported that the crystals of the compound are chiral.

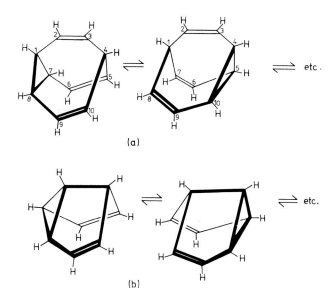

Figure 109. (a) Bullvalene, (b) semibullvalene.

A number of investigators have been involved in the synthesis and studies of 'fluxional' molecules such as 'bullvalene' and 'semibullvalene' (Fig. 109). An examination of the NMR spectrum of bullvalene as a function of increasing temperature reveals the coalescence of the complex spectral lines observed below room temperature to a single signal at high temperatures. At the higher temperature the ten hydrogens become magnetically equivalent due to the rapid isomerization involving both σ- and π-electrons. The isomerization in semibullvalene is so rapid that the equivalence of hydrogens is observed at temperatures considerably below 0 °C. Although the manner in which these structures are written may remind one of the writing of the two Kekulé forms of benzene, the systems are not analogous since the electrons in benzene are uniformly delocalized and the method of writing two formulas is an artificial representation.

The possible synthesis of interlocking carbocyclic rings has been of interest (but not always a serious one) to organic chemists for some time. It was not until 1960, however, that Edel Wasserman at the Bell Telephone Laboratories in New Jersey finally succeeded in preparing the first such 'catenane' (Latin *catena*, chain). The presence of deuterium in the product obtained from an acyloin condensation (Fig. 110) indicated that about 1% of the cyclization was able to take place through the macrocyclic ring (110b). In 1971, I. T. and S. Harrison reported the synthesis of

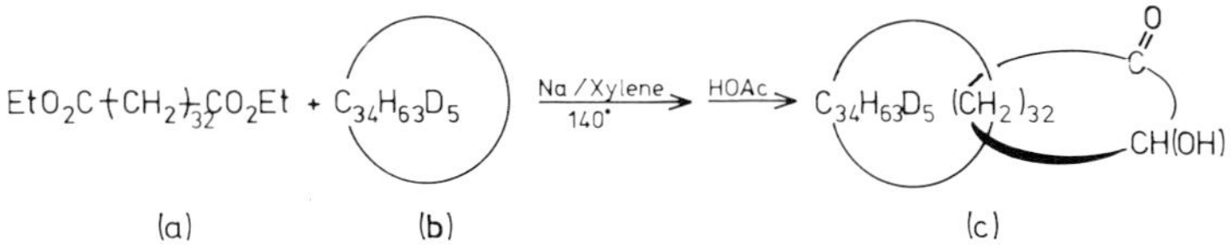

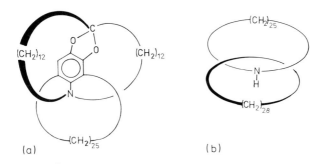

Figure 110. A catenane synthesis.

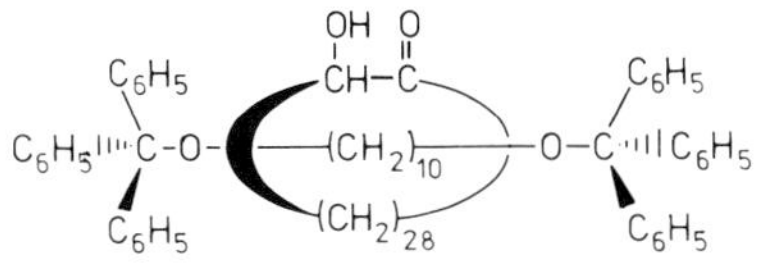

Figure 111.

Figure 112.

a 'hooplane' by a 'threading' chemical reaction. The macrocyclic component (2-hydroxycyclotriacontanone) of the 'hooplane' complex (Fig. 111) was first bound to a polymer support. Treatment with decane-1, 10-diol and chlorotriphenylmethane in pyridine gave about 5% yield of the threaded diether. The first directed synthesis of a catenane (Fig. 112b) (which minimizes cyclizations external to the ring) was reported in 1972 by G. Schill and A. Lüttringhaus. The 'triansa' compound (Fig. 112a) was used as the starting material for the synthesis.

In recent years there have appeared a number of publications directed toward providing a better theoretical basis for understanding the stability and reactivity of molecules of stereochemical interest. We have already mentioned (Chapter 14) Hendrickson's calculations concerning the energy content of the conformations of cyclic compounds. Theoretical calculations have sometimes served to guide the chemists in the synthesis of new compounds which seem unlikely to be capable of existence from the point of view of 'classical' stereochemistry.

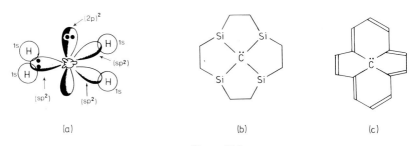

Figure 113.

In 1970, for example, Roald Hoffman, R. Walden and C. F. Wilcox Jr (at Cornell University) initiated a study on the feasibility of the existence of the planar tetravalent carbon atom:

> The tetracoordinate tetrahedral carbon has magnificently served biological systems for millions of years and our imagination for but a century. We here open the problem of stabilizing the tetracoordinate planar carbon.

Their theoretical calculations had shown that a planar carbon might be stable if the carbon was sp^2-hybridized. The four hydrogens might then overlap with the three sp^2 hybrid orbitals (Fig. 113a), two in normal 2-electron 2-center bonds with two hydrogens, the third hybrid orbital might participate in a 2-electron 3-center bond (of the sort found in boron hydrides) with the remaining two hydrogens. The remaining two valence electrons would reside in the non-bonded 2p-orbital. Resonance among these kinds of structures could produce an equivalence of all four hydrogens. The higher energy required to maintain the planar carbon configuration relative to the tetrahedral carbon might be lowered in suitably designed molecules that favored a planar geometry and allowed for the delocalization of the p-electrons. The two compounds (Fig. 113 b, c) were suggested as model compounds, since silicon might be expected to serve not only as an effective σ-donor to the four bonds to carbon, but also use its 3d-orbitals as a π-acceptor for the electrons in carbon's p-orbital. The delocalization of the two p-orbital electrons on the central carbon in 113c to produce a 16π-electron annulene system was thought to be highly favorable to the stabilization of a planar configuration. No report of the synthesis of such compounds has appeared as yet.

In recent years group, graph and set theory have been used to examine systematically the symmetry properties of molecules (discussed earlier in Chapter 8). At Princeton University, Kurt Mislow has adopted this approach to reveal some subtle aspects of the stereochemistry of aryl-methane compounds. By the late 1960s, NMR spectroscopy had indicated

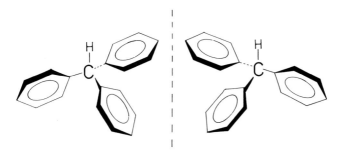

Figure 114. Triphenylmethane enantiomers.

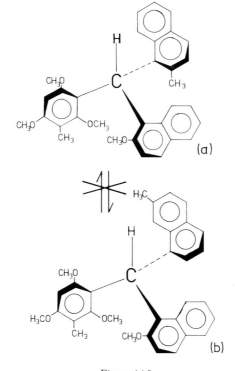

Figure 115.

that the benzene rings in triarylmethanes were tilted in a 'propeller-shape' about the methane carbon (Fig. 114). Since the mirror image of such a propeller-shaped molecule is non-superimposable on itself (that is, the tilt of the rings is in the opposite direction), these compounds should be optically active, if it were not for the fact that rotation about the single bond joining the phenyl rings causes an interconversion of the enantiomers. With suitably bulky *ortho* substituents, this rotation could be prevented, and on the basis of this model we would expect to be able to isolate the stable, optically active enantiomers by conventional resolution techniques. The stereochemistry of the triarylmethanes turns out to be more complicated than is apparent from the above discussion. Mislow was able to show that for triarylmethanes, in which the three aryl groups were different (ArAr'Ar"CH), it was possible to have 32 stereoisomers — even in compounds in which the rotation about the single bond to the aromatic ring is not restricted. The 'excessive' stereoisomerism arises because the rings can only 'flip' in the same direction. Figure 115 indicates the configuration of two diastereomers that were isolated by Mislow. It should be emphasized that the two structures shown in Fig. 115 are not enantiomers, nor are they interconvertible except at high temperatures that allow for the 'reversing' of the propellers. The isolation of these compounds poses problems in stereochemical nomenclature since

> . . . there is no restriction on the torsional angle of any of the individual aryl groups . . . conventional definitions of 'conformer' which are couched in terms of torsional angles about a single bond [Chapter 14] appear to be inadequate to deal with the present case, without modification, since it is difficult to see how any such definition is capable of differentiating between [115 a and b], which are, after all, diastereomeric *conformers.*

Mislow has also suggested the existence of an unusual number of stereoisomers for diarylmethanes, as well as triarylcarbonium ions (ArAr'Ar"C$^{\oplus}$) and triarylamines (ArAr'Ar"N:). It is interesting to note that whether the three bonds to carbon or nitrogen are coplanar does not determine the number of possible stereoisomers. The 'tilt' of the aromatic rings produces the propeller-shaped conformation. The only stable, optically active carbonium ions that have been reported thus far are those compounds whose aryl groups contain *ortho* substituents.

The isolation and study of stable carbonium ions is a rather recent research activity. Since the carbonium ion is normally found only as an unstable intermediate in certain chemical reactions, its geometry could only be inferred from the stereochemistry of the reaction products (Chapter 11). The discovery of 'superacid' media, such as mixtures of antimony

Figure 116. The 1-adamantyl carbenium ion.

pentafluoride with fluorosulfonic acid ($SbF_5 - SO_2 ClF$) or hydrogen fluoride, has made it possible to study the properties of carbonium ions themselves in solution. The examination of these solutions by newer and sophisticated instrumental techniques (^{19}F, ^{13}C NMR spectroscopy, for example) has provided not only conformational details but also evidence for the existence of carbonium ions of controversial structures (such as the 'non-classical' carbonium ion). Even a stable non-planar carbonium ion (Fig. 116) could be shown to exist in superacid media.

George Olah (at Case Western Reserve University), who has been prominent in this area of research, recently suggested a change in the nomenclature: positively charged carbon compounds would be referred to generally as 'carbocations' (analogous to 'carbanions'). The trivalent carbocations should be designated as 'carbenium ions' rather than 'carbonium ions'. The term 'carbonium ion' should be reserved for the designation of penta- (or tetra-) coordinated carbocations such as the methonium ion, $CH_5^{\oplus}$. The methonium ion has been proposed as a transient species formed from the protonation of methane in superacid media. The structure of the methonium ion is still the topic of discussion and experimental study.

The previous discussions have only touched on the significance of the newer instrumental techniques in the development of stereochemical research in the past decade. The function of modern instruments as stereochemical probes has acquired a sensitivity that would be thought impossible a few years ago. For example, Kokke and Oosterhoff used circular dichromism spectra to detect the chirality of the compound shown in Fig. 117. This was the first example of a compound whose chirality was derived from the presence of an isotope other than deuterium (see Chapter 8).

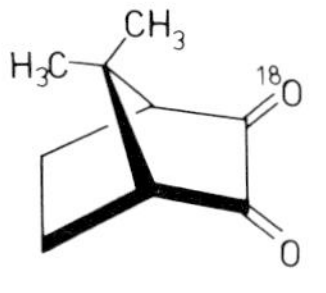

Figure 117.

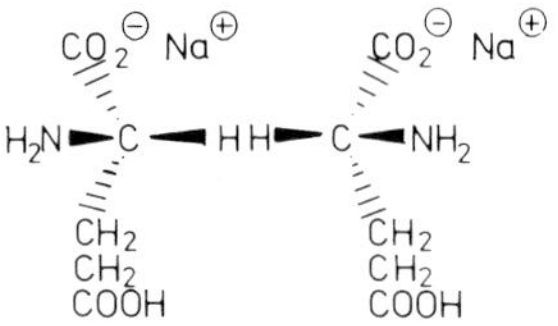

$$CO_2^{\ominus} \ Na^{\oplus} \qquad CO_2^{\ominus} \ Na^{\oplus}$$

$$H_2N \blacktriangleright C \blacktriangleleft H \quad H \blacktriangleright C \blacktriangleleft NH_2$$

$$\begin{array}{cc} CH_2 & CH_2 \\ CH_2 & CH_2 \\ COOH & COOH \end{array}$$

Figure 118. Enantiomers of sodium aspartate.

$$CO_2^{\ominus} \qquad\qquad CO_2^{\ominus}$$

$$\oplus H_3N \blacktriangleright C \blacktriangleleft H \quad H \blacktriangleright C \blacktriangleleft NH_3 \oplus$$

$$\begin{array}{cc} CH_2 & CH_2 \\ C{=}O & C{=}O \\ NH_2 & NH_2 \end{array}$$

Figure 119. Enantiomers of asparagine.

B.   BIOLOGICAL STEREOCHEMISTRY

Perhaps the most significant developments in stereochemistry relate to the research concerned with the properties of biologically important molecules. The chirality associated with many substances isolated from natural sources (for example, the L-amino acids and D-sugars) led chemists to attempt to duplicate such asymmetric transformations (Chapter 15). As the stereochemistry of the synthesis of simpler molecules was better understood, chemists turned their attention to the properties of more complex biological molecules. Living systems seem uniquely designed to distinguish between stereoisomers, sometimes in ways which we may experience ourselves. For example, of the amino acids shown in Fig. 118, the enantiomer shown on the right has almost no taste, while that on the left has a 'meaty' taste. Similarly, the enantiomer shown on the left in Fig. 119 is tasteless, while that on the right is sweet tasting. It is more difficult to detect differences in the odors of enantiomers, but it has been found that the compound shown on the left in Fig. 120 has a spearmint odor while that on the right has a caraway odor.

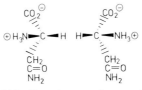

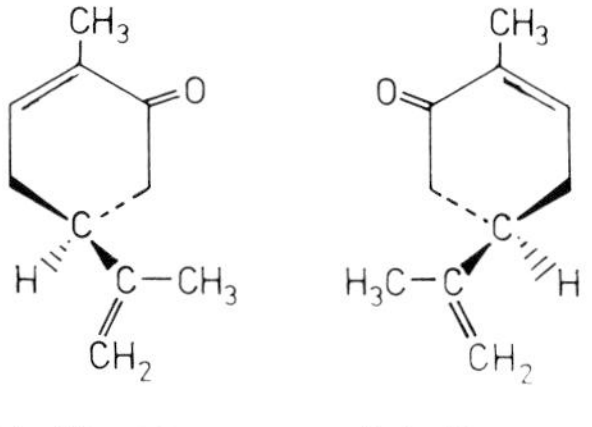

(+) – Carvone   (−) – Carvone

Figure 120.

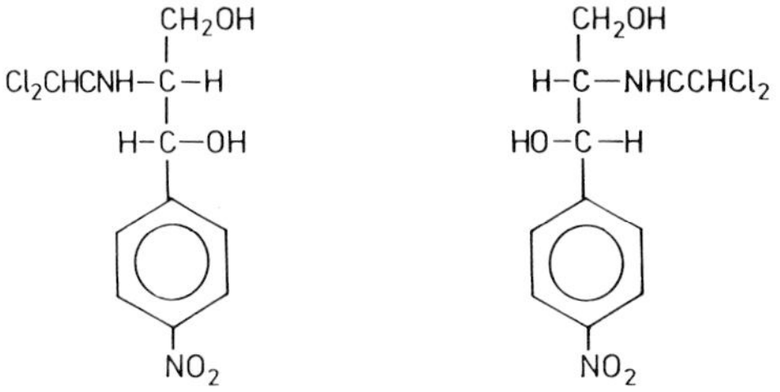

$(-)$-*threo*-Chloramphenicol $(+)$-*threo*-Chloramphenicol

Figure 121.

The influence of the steric configuration on the activity of drugs has been known for some time. An example from the more recent literature is illustrated in Fig. 121. Whereas $(-)$-*threo*-chloramphenicol is a potent antibactericide, the $(+)$-*threo*-enantiomer shows negligible activity. The *erythro* enantiomers also show little or no antibacterial activity. The observation that stereoisomers may differ in their biological activity can certainly date back to Pasteur's study of the action of moulds on racemic tartrate solutions (Chapter 4). An early example of the differing pharmacological activity of enantiomeric compounds was provided by Arthur Cushing in the first decades of the 20th century. It will be possible to give only a brief review of some of the research concerned with the stereochemistry of drug action.

A number of theories have been proposed over the years to explain the activity of drugs, and many of them relate to the stereospecificity of the enzymes found in living systems. The earliest general theory to account for the stereospecificity of enzyme activity was the 'lock-and-key' proposal of Emil Fischer (Chapter 15). In 1933 Leslie Easson and Edgar Stedman, at the University of Edinburgh, proposed that for a drug to produce its maximum physiological effect, it must be attached to the receptor site [the enzyme surface perhaps] at three points. The requirement for a specific three-point attachment allows the receptor site to distinguish between enantiomers. That enzymes were capable of an even more subtle kind of stereochemical discrimination was not apparent until the 1930s with the more extensive use of isotopic labeling to study the pathways of enzymatic reactions. In 1941 H. G. Wood, C. H. Werkman, A. Hemingway and A. O. Nier using $^{13}CO_2$, and E. A. Evans and L. Slatin using $^{11}CO_2$, reported that the α-ketoglutaric acid formed in what is now known as the 'tricarboxylic acid cycle' (or 'citric acid' or 'Krebs' cycle) in the presence of labeled CO_2 and pyruvic acid was labeled solely in the carboxyl group adjacent to the carbonyl group. Oxaloacetic acid was found to be an intermediate in this reaction:

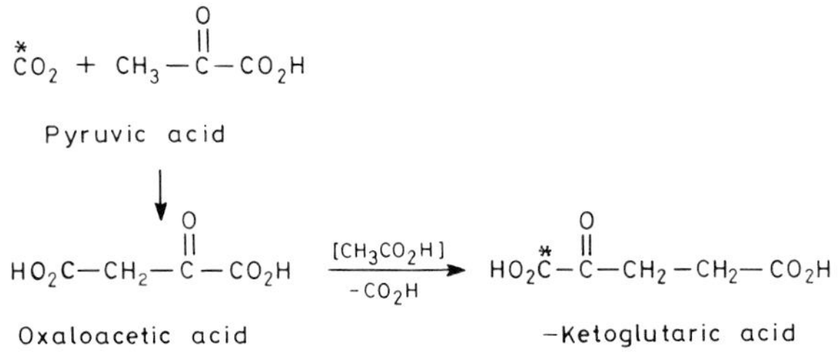

$$\overset{*}{C}O_2 \; + \; CH_3 - \overset{\overset{\displaystyle O}{\|}}{C} - CO_2H$$

Pyruvic acid

$$HO_2C - CH_2 - \overset{\overset{\displaystyle O}{\|}}{C} - CO_2H \xrightarrow[-CO_2H]{[CH_3CO_2H]} HO_2\overset{*}{C} - \overset{\overset{\displaystyle O}{\|}}{C} - CH_2 - CH_2 - CO_2H$$

Oxaloacetic acid   –Ketoglutaric acid

At about this time, Hans Krebs had postulated that a series of reactions were involved in the formation of α-ketoglutaric acid. He predicted that the isotopically labeled CO_2, fixed in oxaloacetic acid, would appear in the α-ketoglutaric acid, but that the labeling should be distributed equally between the two carboxyl groups. Krebs was led to this conclusion because of what he assumed was the 'symmetry' of the citric acid postulated as an intermediate. The formation of α-ketoglutaric acid involved the decarboxylation and oxidation of isocitric acid, another postulated intermediate (Fig. 122). Since the above scheme suggested that there should

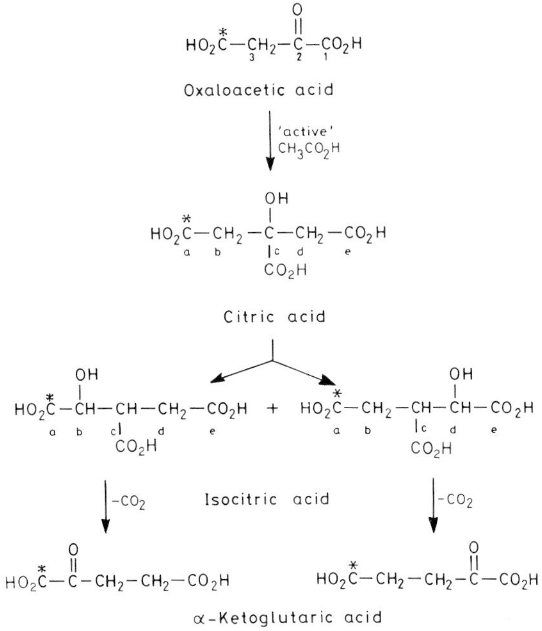

$$HO_2\overset{*}{\underset{3}{C}} - CH_2 - \overset{\overset{\displaystyle O}{\|}}{\underset{2}{C}} - \underset{1}{CO_2H}$$

Oxaloacetic acid

'active'
CH_3CO_2H

$$HO_2\overset{*}{\underset{a}{C}} - \underset{b}{CH_2} - \overset{\overset{\displaystyle OH}{|}}{\underset{\underset{CO_2H}{|c}}{C}} - \underset{d}{CH_2} - \underset{e}{CO_2H}$$

Citric acid

$$HO_2\overset{*}{\underset{a}{C}} - \overset{\overset{\displaystyle OH}{|}}{\underset{b}{CH}} - \underset{\underset{CO_2H}{c|}}{CH} - \underset{d}{CH_2} - \underset{e}{CO_2H} \quad + \quad HO_2\overset{*}{\underset{a}{C}} - \underset{b}{CH_2} - \overset{\overset{\displaystyle OH}{|}}{\underset{\underset{CO_2H}{|c}}{CH}} - \underset{d}{CH} - \underset{e}{CO_2H}$$

$-CO_2$ Isocitric acid $-CO_2$

$$HO_2\overset{*}{\underset{}{C}} - \overset{\overset{\displaystyle O}{\|}}{C} - CH_2 - CH_2 - CO_2H \qquad HO_2\overset{*}{C} - CH_2 - CH_2 - \overset{\overset{\displaystyle O}{\|}}{C} - CO_2H$$

α-Ketoglutaric acid

Figure 122. H. Krebs' proposal for the formation of α-ketoglutaric acid from oxaloacetic acid.

be no difference between the two methylene groups (b) and (d) that are ultimately oxidized to the α-keto group, the original label should be distributed equally between the two carboxyl groups. This prediction was contradicted by the labeling experiments of Wood *et al.*, and Evans and Slotin. They therefore concluded that citric acid could not be an intermediate involved in the formation of α-ketoglutaric acid.

No solution to this problem was suggested until 1948 when A. G. Ogston, at Oxford University, pointed out that the experimental data did not preclude citric acid as an intermediate. Ogston stated:

> On the contrary, it is possible that an asymmetric enzyme which attacks a symmetrical compound can distinguish between its identical groups.

Ogston's argument was based on the assumption that a 'three-point combination occurs between the symmetrical substrate and the enzyme'. Although he mentioned that 'such combination is, of course, necessary wherever a single optical antipode is formed enzymatically from an inactive precursor', he did not specifically mention Easson and Stedman's earlier proposal. The manner in which an enzyme might distinguish between two apparently identical groups is illustrated in Fig. 123. (The arguments presented here are similar to those given by Ogston, who used the reaction of aminomalonic acid as his example.) We first imagine that the enzyme contains three 'receptor' sites for the binding of the substrate. If we designate site 'B' as the active site responsible for the oxidation of the methylene group to the α-keto function, then it should be apparent that the carbon numbered 3, rather than 1, is oxidized. It is not possible to oxidize carbon number 1, since the rotation of 120° about the C–OH bond that places that group at site B, places non-binding groups ($CH_2 CO_2 H$ and $CO_2 H$, respectively) in sites C and A. It is clear, then, that the carbonyl group is introduced only on the carbon adjacent to the labeled carboxyl in oxaloacetic acid (Fig. 122).

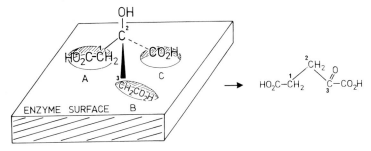

Figure 123. Illustration of Ogston's 'three-point' receptor site hypothesis.

Many biochemists had .difficulty in grasping the meaning of Ogston's paper. In a recent book concerned with the applications of stereochemistry to biochemistry, William Alworth has reviewed some of the difficulties associated with Ogston's observations:

The differentiation of the paired carboxymethyl ($-CH_2 CO_2 H$) groupings of citric acid in the reactions of the TCA cycle has come to be known as the Ogston effect . . . Ogston's initial proposal stated that enzymes can distinguish between the 'identical' groups of a 'symmetrical' compound. Whatever Ogston's intended meaning, this statement led to an unfortunate amount of confusion among biochemists. The Ogston effect was interpreted to mean that identical groups on symmetrical molecules were distinguished by enzymes in some mysterious biological manner involving three points on an enzyme surface. The three-point-attachment concept was perverted to mean that the formation of citrate and its conversion to isocitrate occurred via totally enzyme bound intermediates; that is, the free 'symmetrical' citrate molecule was never actually present in the reaction media. In actual fact, Potter and Heidelberger demonstrated in 1949 that biologically formed citrate that was isolated, purified, and then added to a second enzyme preparation still resulted in the formation of specifically labeled α-ketoglutarate.

The misconceptions regarding this aspect of enzyme stereospecificity can largely be traced to an imprecise use of the term 'symmetrical'. A careful, critical examination of the representation in [Fig. 123] should lead to the observation that the three-point attachment to the enzyme surface is *not inducing asymmetry* into the citric acid molecule; rather, the enzymatic attachment pictured just illustrates schematically how *an inherent lack of symmetry in the citric acid may be recognized and exploited by an enzyme.*

The above passage has been quoted at some length since even at this date, many persons have difficulty in seeing that '. . . the chemically like, paired carboxymethyl groups are *inherently nonequivalent*, and therefore it is possible to differentiate them'.

It is not possible at this time to explore this problem further. In modern terminology, such pairs of chemically like groups are said to be *enantiotopic* if their separate replacement by another group (which is different from the others already present) gives rise to a pair of *enantiomers*; and *diastereotopic* if diastereoisomers are obtained upon replacement (see Glossary).

In a more general theory of enzyme activity, the receptor sites con-
stitute what are termed the 'active site' on the enzyme surface. In this
theory, the enzyme exhibits its extraordinary catalytic activity because
the substrate fits into a 'pocket' in the enzyme in such a way as to
maximize the chemical activity of the reacting functional groups in the
enzyme. An enzyme that has been much studied is the hydrolytic enzyme
α-chymotrypsin, a protein composed of 245 amino acids. This enzyme
cleaves the amide bonds between certain amino acids in polypeptides. By
the 1960s the results of numerous ingenious chemical studies had indicated that the catalytic activity of the enzyme was associated with three
amino acids in the active site. These were histidine, aspartic acid and
serine which were located as amino acids number 57, 102 and 195 in the
protein chain. The mechanism by which these amino acids might function
in the cleavage of a peptide bond is summarized in Fig. 124. The hydro-
lytic cleavage of a peptide linkage $(R'-\overset{O}{\overset{\|}{C}}-NH-R+H_2O \rightarrow R'-\overset{O}{\overset{\|}{C}}-OH+$
$H_2NR)$ involves first the acylation of the enzyme (on the hydroxymethyl
group of serine-195) followed by the hydrolytic deacylation to regenerate
the enzyme. Serine-195 thus serves as a nucleophile, while aspartic acid
and histidine function as general acid or base catalysts. The substrate presumably finds itself in the proper orientation in the active site so that

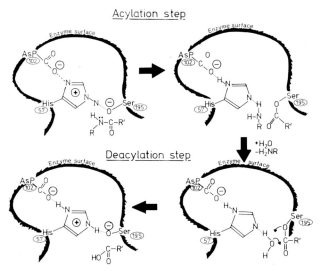

Figure 124. Mechanism proposed for the hydrolysis of a peptide bond at
the active site in α-chymotrypsin.

these functional groups can exert their maximum catalytic activity. The mechanistic studies could only suggest that such an orientation was required. In the 1960s, high resolution X-ray diffraction spectroscopy indicated that chymotrypsin was a globular protein that was coiled in such a way that its more hydrophobic (non-polar) parts of the amino acids were turned inward, away from water, to permit a maximum amount of intramolecular hydrogen bonding. The enzyme also contained a hydrophobic 'pocket' in which the non-polar parts of substrate molecules might fit. The three key amino acids were found located in the surface of this 'pocket'. How this 'pocket' functioned as the active site could also be shown by X-ray spectroscopy of an enzyme-substrate complex. Since this study was not possible for a normal substrate, because of the rapid cleavage reaction, the structure of the enzyme complex with an enzyme inhibitor, N-formyl-L-tryptophane, was determined. T. A. Steitz, R. M. Henderson and D. M. Blow (at the Medical Research Council Laboratory at Cambridge) were able to demonstrate clearly that the aromatic ring rested in the hydrophobic pocket so as to orient the amide function so that it was in reasonably close proximity to the three amino acids (Fig. 125). An even closer fit was expected for normal substrate molecules. It is interesting to note that the fit of the substrate into the enzyme pocket is achieved with little conformational change in the enzyme. In other enzymes, such as carboxypeptidase-A, the insertion of a substrate produces a number of significant conformational changes in the active site.

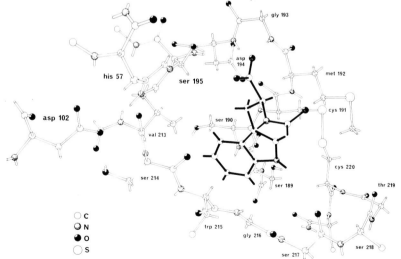

Figure 125. Partial structure of the active site of α-chymotrypsin containing N-formyl-L-tryptophane (atoms joined by filled-in bonds).

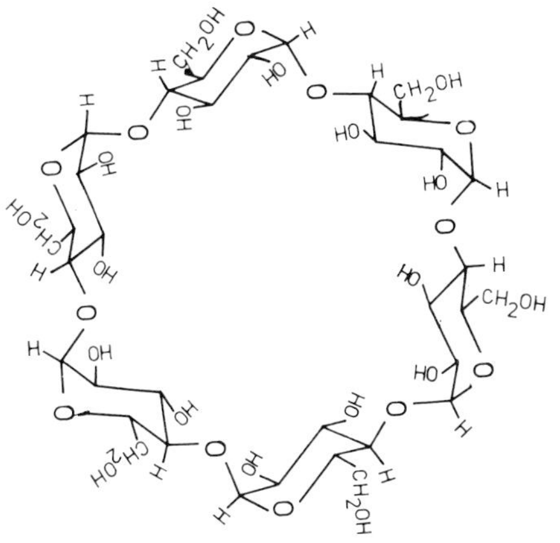

Figure 126. The cyclodextrin: cyclohexaamylose.

When it was realized that the remarkable catalytic activity and specificity of enzymes could be understood in terms of relatively straightforward stereochemical and mechanistic concepts, it was not long before chemists undertook the synthesis of simpler organic catalysts that might exhibit similar activity. One of the first class of compounds that were studied for this reason were the cyclodextrins, which are cyclic carbohydrates composed of six to eight glucose units linked together by α-1,4-glycosidic bonds (Fig. 126). In a model of cyclohexaamylose, the ether oxygens and hydrogens turn inwards to produce a somewhat hydrophobic hole, which is sufficiently large to accommodate an aromatic ring; the secondary hydroxyl groups turn outward. In the 1960s, Myron Bender (at Northwestern University), as well as other investigators, found that the cyclodextrins exhibit a remarkable catalytic activity in the hydrolysis of certain aromatic esters. For example, *m-tert*-butylphenyl acetate underwent hydrolysis 200 times more rapidly in the presence of cyclohexaamylose than did phenyl acetate. The unusual effect of the *tert*-butyl substituent was shown with models to be caused by the fit of the alkyl group into the cycloamylose cavity in such a way that the ester function was placed in close proximity to one of the secondary hydroxyl groups on the exterior of the ring. In basic solution, the alkoxide ion of the secondary hydroxyl group could function as a nucleophilic species in a manner analogous to the hydroxymethyl group of serine-195 in chymotrypsin (Fig. 124).

In 1972, E. T. Kaiser (at the University of Chicago) observed that the (+)-enantiomer of the compound shown in Fig. 127 hydrolyzed more rapidly than the (−)-enantiomer in the presence of cyclohexaamylose. A similar enantiomeric specificity could be demonstrated for α-chymotrypsin. The cause of the specificity is again found in the fit of the

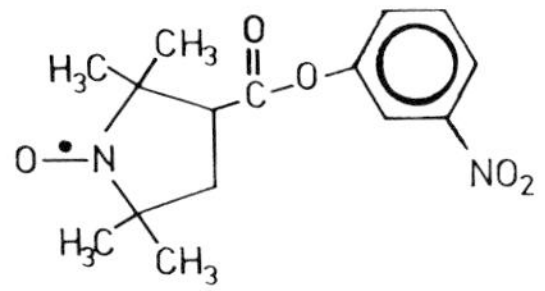

Figure 127.

aromatic ring which produces a proper orientation of the ester function toward the nucleophilic hydroxyl groups in only one of the enantiomers.

Recently there has been considerable interest in the properties of a series of cyclic polyethers – often referred to as 'crown' ethers. These crown ethers exhibit a remarkable ability selectively to complex cations. In the cyclic structure, the ether functions are oriented within the ring, while the more non-polar or hydrophobic parts of the structure form the external shell. The crown ether solvates the cations so well that it is possible to use them, for example, to dissolve potassium permanganate in benzene. The optically active form of the cyclic polyether shown in Fig. 128 could be used to separate (S)-valine from a racemic mixture. The crown ether:(S)-valine complex could be extracted from an aqueous acetic acid solution by chloroform. The ammonium ion of valine is bound in the center of the cavity by hydrogen bonds and ionic bonds to the ether functions and one of the carboxyl ions. More efficient methods of resolution using these crown ethers will undoubtedly be reported in the near future. If the crown ether were bound to a polymer support, it would then be possible to separate enantiomers in a continuous flow process. If catalytic functional groups were also introduced at the entrance to

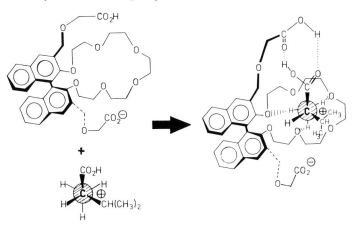

Figure 128. Stereospecific binding of (S)-valine, $(CH_3)_2 CH\, CH(NH_3^+)CO_2 H$, by a chiral 'crown' ether.

the cavities, it should be possible to produce 'model' enzymes having specificities and activities unknown in natural enzymes.

The above discussions should indicate how important an understanding of stereochemical concepts has been to the recent development of biochemistry. It is not possible in this book to cover all of the developments in this field. It would be worthwhile, however, to return to discuss those areas of the application of stereochemistry to biochemistry to which John Cornforth has contributed.

Cornforth received the Nobel Prize in Chemistry in 1974 for his contributions to the stereochemistry of squalene biosynthesis. Cornforth's research, however, illustrates the more subtle aspects of stereochemistry that is only now beginning to be appreciated by biochemists and organic chemists. The observations made by Ogston in 1948 illustrate what can

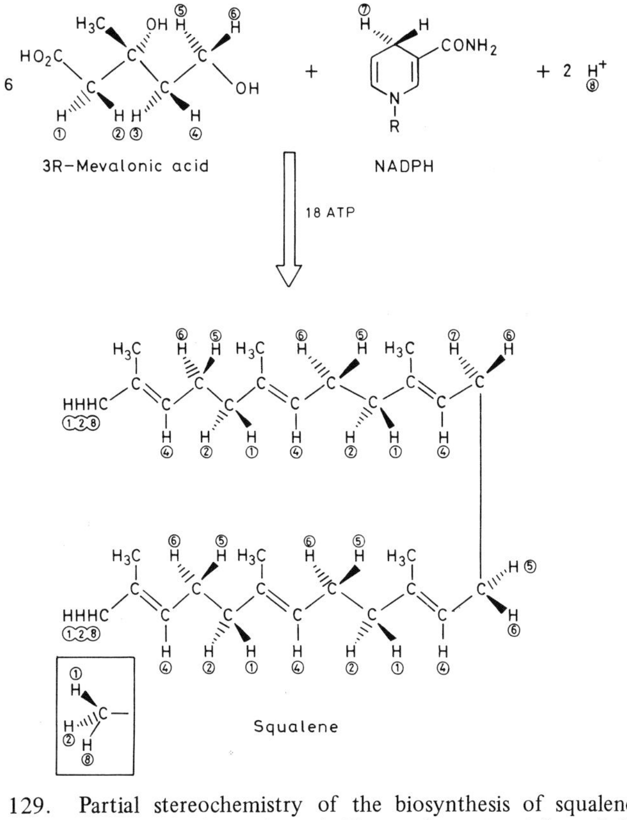

Figure 129.   Partial stereochemistry of the biosynthesis of squalene. (The fate of hydrogens 1–8 is indicated. The configurational fate of the three hydrogens in the terminal methyl groups was determined in 1972.)

be described as *hidden stereochemistry*. For example, in citric acid it is not apparent that the enzyme can distinguish between the two carboxyl-methyl groups ($-CH_2 CO_2 H$). It is only possible to see this stereo-chemistry by the use of isotopic labeling. The synthesis of squalene from 3R-mevalonic acid (Fig. 129) involves five intermediate steps (not shown in the figure). Cornforth and his associate, George Popjak, established the stereospecificity of 14 points of 'stereochemical ambiguity' in the squalene biosynthesis. It is beyond the scope of this discussion to discuss all of these points. The determination of the fate of individual hydrogens (1–8 in Fig. 129) by Cornforth and Popjak involved the use of deuterium and tritium labeling. Although the fate of labeled hydrogens could be determined by mass spectrographic analysis, the establishment of the stereochemistry involved required analytical procedures that were only developed in the 1960s. For example, their work coincided with the development of polarimeters sensitive enough to measure the optical activity due solely to the substitution of hydrogen by deuterium in samples of a few milligrams. As Cornforth described the work in his Nobel address:

> Thus the work required in unusual measure the harmonious blending of stereospecific synthesis, isotopic labeling, enzymo-logy, chemical degradation on the centigram scale, and sensitive physical methods of analysis into a single experimental se-quence. In the end, we succeeded in demonstrating stereo-specificity for all but one of the enzymic steps then known for squalene biosynthesis.

One step that remained unsolved was the isomerization of isopentenyl pyrophosphate into dimethylallyl pyrophosphate. The manner in which the stereochemistry of this reaction was elucidated is discussed in Chapter 1. Central to the solution of the problem was the synthesis of a chiral methyl group – using all three isotopes of hydrogen: protium, deuterium, and tritium. This seemed an impossible task at the time:

> Chiral methyl groups were unknown at the time, and it was not obvious how their absolute configuration could be determined: optical rotation was an unlikely candidate for measurement since a substance having tritium in atomic proportion (instead of the small labeling concentration) would have a specific radio-activity of some 30,000 curies per mole, and the rotary power would probably be so small as to require large specimens having this order of radioactivity.

The solution to the problem was provided by the suggestion of Herman Eggerer, who predicted that an isotope effect should be observed in the displacement of hydrogen from a chiral methyl group in the reaction of acetyl-coenzyme A and glyoxylic acid to produce malic acid. The results of this study (discussed in Chapter 1 and in the reprint of the 1969 paper) were then used to solve the remaining stereochemical uncertainty in the squalene biosynthesis. The configurational fate of the three hydrogens (1, 2 and 8 in Fig. 129) in the two terminal methyl groups of squalene was determined in 1972.

The studies of Cornforth on the chiral methyl group illustrate an aspect of enzyme chemistry that is not often appreciated. This is perhaps best summarized by Cornforth himself:

> Our adventure with the chiral methyl group reinforced the conviction that stereospecificity is something not just incidental, but essential, to enzymic catalysis. Life does depend on accurate replication of molecules, and its complexity often requires that an enzyme shall accept one molecular species or type and transform it to equally specific products. But the hidden specificity that we have helped to reveal goes much further than this. An enzyme must, it seems, catalyze strictly stereospecific reactions even when this specificity is not required by the structural relation of substrate to product. Indeed, many examples are now available in which an enzyme can accept more than one molecular species as substrate but still transforms each of them with absolute, though hidden, control of the stereochemistry of reaction.

Although this chapter has emphasized the importance of stereochemistry to the area of biochemistry, other areas are also worthy of study. For example, the dramatic change that took place in the field of polymer chemistry in the past two decades is due in large part to G. Natta's discovery in the 1950s of the conditions necessary to produce stereoregulated polymers. The ability to synthesize stereospecifically *isotactic, syndiotactic* or *atactic* polymers has had an enormous theoretical and commercial impact on the direction of polymer research. Hopefully, another book in this series will discuss these developments in more detail. For the moment, we will have to be content with the limits imposed by this book.

Appendix A

Chronology of events and publications in the history of stereochemistry

(Titles of some key papers indicated in quotes)

1690	C. Huygens	Study of crystallography and polarized light
1784	A. R. J. Haüy	Crystallographic studies; hemihedrism of quartz crystals noted
1803	J. Dalton	Atomic theory
1809	E. L. Malus	Theory of polarized light
1811	A. Avogadro	Atomic-molecular hypothesis
1813	J. B. Biot	Optical rotation of plane polarized light by quartz crystals
1815	J. B. Biot	Optical activity of some liquids from natural sources
1818	J. B. Biot	Optical activity of sugar solution; inverse square law
1820	J. F. W. Herschel	Sign of optical rotation and hemihedrism of quartz crystals
1820	C. Kestner	Obtains isomeric form of tartaric acid (racemic form)
1821	A. Fresnel	Theoretical studies of polarized light
1832	J. Berzelius	Concept of isomerism
1832	J. B. Biot	Optical activity of tartaric acid
1838	J. B. Biot	Optical inactivity of racemic tartaric acid
1844	E. Mitscherlich	Reports that crystals of the sodium—ammonium salts of tartaric acid and paratartaric acid have identical properties
1848	L. Pasteur	Separation of enantiomeric crystals of sodium—ammonium paratartrate
1853	L. Pasteur	Chemical synthesis of racemic and *meso*-tartaric acid
1858	L. Pasteur	(a) Chemical resolution of racemic tartaric acid via recrystallization of alkaloid salts
		(b) Biochemical 'resolution' of racemic tartaric acid using *Penicillum glaucum* mould
1858	Formulation of concept of tetravalency and chain-linking capacity of carbon:	
	A. Kekulé	'The Constitution and Metamorphoses of Chemical Compounds and the Chemical Nature of Carbon'
	A. S. Couper	'On a New Chemical Theory'
1858	A. Kekulé	Use of 'roll' formula and graphic formula
1860	S. Cannizzaro	Conference at Karlsruhe: atomic weights

1861	A. M. Butlerov	Chemical structure
1862	A. M. Butlerov	Valencies in carbon tetrahedrally oriented
1863	J. Wislicenus	Lactic acid isomers: arrangements of atoms in space
1864	A. Crum Brown	'Constitutional' formulae
1865	A. Kekulé	Hexagon structure for benzene
1865	A. Hofmann	'Croquet ball' models
1867	A. Kekulé	Tetrahedral carbon models
1867		Commercial sale of 'glyptic' models
1869	W. Koerner	Benzene model (tetrahedral carbon)
1869	E. Paternò	$C_2H_4Br_2$ isomers (tetrahedral carbon)
1869	D. I. Mendeleev	Periodic table
1873	J. Wislicenus	Lactic acid isomers: arrangement of atoms in space
1874	J. H. van't Hoff	'A Suggestion Looking into the Extension into Space of the Structural Formulas at Present Used in Chemistry. And a Note upon the Relation between the Optical Activity and the Chemical Constitution of Organic Compounds' (Dutch pamphlet)
1874	J. A. Le Bel	'On the Relations Which Exist between the Atomic Formula of Organic Compounds and the Rotatory Power of their Solutions' (French paper)
1875	J. H. van't Hoff	'La Chimie dans l'Espace' (French pamphlet)
1876	V. Meyer	Difficulty of synthesis of small rings
1877	J. H. van't Hoff	'Die Lagerung der Atome in Raume'
1883– 1885	W.H. Perkin	Synthesis of derivatives of cyclobutane, cyclopropane and cyclopentane
1885	A. von Baeyer	Tension theory
1887	J. Wislicenus	'The Space Arrangements of the Atoms in Organic Molecules and the Resulting Geometrical Isomerism in Unsaturated Compounds'
1888	F. Kehrmann	Steric hindrance in o-substituted quinones
1890	A. Hantzch and A. Werner	'On the Spatial Arrangement of Atoms in Nitrogen Containing Molecules'
1890	H. Sachse	'On the Geometrical Isomers of Hexamethylene Derivatives'
1890	C. A. Bischoff	'Limitation of the Free Rotation of Single Bond Carbon Atoms'
1891	E. Fischer	'Conformation of Grape Sugar and its Isomerides'
1891	J. A. Le Bel	Optically active $R_1R_2R_3R_4N^+X^-$
1893	A. Werner	'Contribution to the Constitution of Inorganic Compounds'
1894	V. Meyer	Steric hindrance; Law of Esterification
1896	P. Walden	Chemical interconversion of enantiomers
1899	W. Marckwald and A. McKenzie	Kinetic method of resolution
1899	W. J. Pope and S. J. Peachey	Optically active $R_1R_2R_3R_4N^+X^-$

1900	W. J. Pope and S. J. Peachey, S. Smiles	Optically active $R_1 R_2 R_3 S^+ X^-$
1900	W. J. Pope and S. J. Peachey	Optically active $R_1 R_2 R_3 Sn^+ X^-$
1902	W. J. Pope and A. Neville	Optically active $R_1 R_2 R_3 Se^+ X^-$
1904	W. Marckwald, A. McKenzie	Asymmetric synthesis
1906	M. A. Rosanoff	Relative configuration of sugars
1907	F. Käufler	'Folded' benzidine structure
1907	F. S. Kipping	Diastereomers of $(R'R''R'''Si)_2 O$
1908	J. Meisenheimer	Optically active $R_1 R_2 R_3 N = O$
1909	W. Perkin, W. J. Pope, O. Wallach	Optically active 4-methylcyclohexylideneacetic acid
1910	F. Challenger and F. S. Kipping	Optically active $R_1 R_2 R_3 R_4 Si$
1911	J. Meisenheimer and L. Lichtenstadt	Optically active $R_1 R_2 R_3 P = O$
1911	S. J. Kipping and F. Challenger	Optically active $(R_1 O)(R_2 O)(RNH)P = O$
1911	W. H. Mills and A. M. Bain	Optically active oxime
1911	A. Werner	'Toward an Understanding of the Asymmetric Cobalt Atom'
1912	J. Cain, A. Coulthard and F. Micklewait	Isomeric 3,3'-dinitrobenzidine
1913	J. Bredt	Bredt's rule
1914	A. Werner	'On Optical Activity Among Carbon-free Compounds'
1915	J. F. Thorpe and C. K. Ingold	Valency deflection hypothesis
1916	G. N. Lewis	'The Atom and the Molecule'
1918	E. Mohr	'The Baeyer Tension Theory and the Structure of Diamond'
1921	H. G. Derx	Conformation of cyclic diols
1921	G. J. Burrows and E. E. Turner	Optically active $R_1 R_2 R_3 As^+ X^-$
1921	J. Kenner and W. F. Stubbins	Optically active 6,6'-dinitrodiphenic acid
1923	W. Hückel, A. Windaus and G. Reverez	Synthesis of isomeric (multiplanar) bicyclic compounds, decalin
1923	G. N. Lewis	'Valence and the Structure of Atoms and Molecules'
1924	P. H. Hermans	Conformation and reactivity of cyclic and acyclic diols

1924	J. Böeseken and J. Meulenhoff, A. Rosenheim and Vermehren	Optically active esters of boric acid
1925	H. Phillips	Optically active sulfinic ester
1925	W. H. Mills and R. Raper	Optically active $R_1 R_2 R_3 As=S$
1925	W. H. Mills and E. H. Warren	Optically active spirane (nitrogen in center)
1925	H. Phillips	Walden inversion (experimental determination of inversion reaction)
1926	H. Phillips	Optically active sulfone
1926	P. W. B. Harrison, J. Kenyon and H. Phillips	Optically active sulfoxide
1926	W. H. Mills and R. A. Gotts	Optically active beryllium compounds
1926	E. E. Turner and R. J. W. LeFevre, F. Bell and J. Kenyon, W. H. Mills	Cause of rotational barrier in biphenyl compounds
1926	S. B. Hendricks and C. Bilicke	X-ray spectroscopy of β-benzenehexachloride
1926	L. Ružička	Synthesis of macrocyclic rings
1927	A. Weissberger, J. Clark, L. W. Pickett, J. Meisenheimer	Dipole moment studies on biphenyl compounds
1929	H. N. Haworth	'Conformation' of pyranose ring
1929	W. Kuhn	Absolute asymmetric synthesis
1929	T. M. Lowry and F. L. Gilbert	Optically active $R_1 R_2 R_3 Te^+ X^-$
1930	L. Fieser	Stereochemistry of steroids
1930	A. Weissberger, K. L. Wolfe	Dipole moment studies of acyclic diastereomers
1931	Schwartz and Lewinsohn	Optically active $R_1 R_2 R_3 Ge^+ X^-$
1931	Reihlen and Huhn	Optically active Pd and Pt complexes
1932	R. Robinson	Electronic theory of chemical reactions
1932	S. E. Janson and N. J. Pope	Optically active spirane
1932	H. Eyring	Calculation of rotational barrier of ethane (0.3 kcal/mole)
1934	H. A. Stuart	Space-filling molecular models
1934	C. K. Ingold	Electronic theory of chemical reactions

1935	W. H. Mills, P. Maitland, E. P. Kohler, J. T. Walker and M. Tishler	Optically active allenes
1936	K.W.F. Kohlrausch	Raman spectral studies of cyclic compounds
1936	J. D. Kemp and K. Pitzer	Calculation of rotational barrier of ethane (3 kcal/mole)
1940	R. Reeves	Conformation of glycosides
1940	H. C. Brown	Investigation of steric effects
1943	O. Hassel	'The Cyclohexane Problem'
1946	O. Bastiansen and O. Hassel	Conformation of decalin
1947	D. H. R. Barton	Calculation of conformational stabilities of cyclohexane and decalin
1950	V. Prelog	Conformation of medium-sized rings
1950	D. H. R. Barton	'Conformation of the Steroid Nucleus'
1951	J. M. Bijvoet	Experimental determination of absolute configuration of $(+)$-tartaric acid
1951	R. S. Cahn and C.K. Ingold, V. Prelog	(R)- and (S)- convention for designation of absolute configuration
1951	L. Pauling, R. B. Corey and H. R. Branson	α-Helix structure of proteins
1952	D. J. Cram and F. A. Abd Elhafez	Rule for asymmetric synthesis
1953	V. Prelog	Rule for asymmetric synthesis
1953	J. D. Watson and F. H. C. Crick	Double helix structure of DNA
1954– 1955	E. L. Eliel and C. A. Lukach; S. Winstein and N. J. Holness	Quantitative expression of conformational reactivity
1955	G. Natta	Stereospecific polymerization
1958	M. F. Perutz and J. C. Kendrew	X-ray structural studies of myoglobin
1965	R. B. Woodward and R. Hoffmann	Conservation of orbital symmetry

Glossary

Absolute configuration. See *Configuration.*

Achiral molecule. A molecule that is not chiral.

Anchimeric assistance. See *Neighboring group participation.*

Angle strain (Baeyer strain). Instability produced by the deviation of the bond angle from the normal bond angle.

Antimers. See *Enantiomers.*

Asymmetric carbon atom. A carbon atom which contains four different substituents (see *Chiral center*).

Asymmetric molecule. A molecule devoid of all symmetry elements (see under *Symmetry*).

Asymmetric synthesis. A reaction in which new chiral centers or groupings are produced in unequal amounts. *Absolute asymmetric synthesis* is concerned with such syntheses that arise from the operation of dissymmetric physical influences, such as circularly polarized light.

Atactic polymer. See *Stereoregular polymer.*

Atropoisomeric molecules. Separable conformational isomers (usually because of presence of high rotational barriers).

Axial positions. Those groups that are oriented perpendicular to the average plane of the carbon atoms in the chair form of cyclohexane (that is, they are essentially parallel to the C_3 axis).

Axis of symmetry. See *Symmetry.*

Baeyer strain. See *Angle strain.*

Baeyer strain theory. See *Strain theory.*

Biot's law. See *Specific rotation.*

Carbenium ion (carbonium ion). A positively charged carbon atom which contains three substituents. A term that has been suggested to replace *carbonium ion.*

Center of symmetry. See *Symmetry.*

Chiral center (asymmetric center). Usually refers to an atom (such as the *asymmetric carbon atom*) to which are attached four different groups. The center of chirality need not always coincide with an atom. (For example, adamantane possessing different groups on carbons 1, 3, 5 and 7 is chiral, but the chiral center is in the space at the center of the molecule.) When substituents are dissymmetrically substituted about an axis in a molecule (e.g. in the biphenyls, allenes etc.) the axis is known as the *chiral axis*. When different groups are located on the two sides of a plane passed through a molecule, the plane is termed the *chiral plane*. (For example, *trans*-cyclooctene, the ansa-compounds are chiral molecules due to the presence of a chiral plane.) Hexahelicene is an example of a compound that possesses *helical chirality*.

Chirality (dissymmetry). When an object or molecule possesses the property of 'handedness', that is, its mirror image is non-superimposable upon itself.

Circularly polarized light. See under *Plane polarized light*.

Cis—trans isomers (geometrical isomers). Stereoisomers that cannot be interconverted because of the energetic barrier to rotation about a carbon—carbon double bond (or the breaking of a single bond in a cyclic compound). Examples: *cis*- and *trans*-2-butene, *cis*- and *trans*-1,2-dimethylcyclopropane.

Configuration. The description of the arrangement of atoms in space. *Relative configuration* refers to a particular configuration that is related (usually by chemical methods) to the arbitrarily selected configuration of a reference compound. *Absolute configuration* is the actual arrangement of atoms in space. A convention may be employed (e.g. the *R* and *S* convention) that describes a particular configuration; the use of this convention does not, however, establish the actual arrangement of the atoms in space.

Conformation. Generally refers to the various configurations of a molecule that are interconvertible by rotation about single bonds.

Conformational analysis. The analysis of the physical and chemical properties of a compound in terms of its molecular conformations.

Conformers (conformational isomers, rotamers). Isomers that may be interconverted by torsion (rotation) about a single bond. The term is sometimes used more restrictively to designate those conformations of minimum energy.

Constitutional formula. A formula that indicates the elementary constituents.

Diastereomers (diastereoisomers). Stereoisomers that are not enantiomers. Originally this term was used with reference to a series of stereoisomers containing two or more asymmetric carbon atoms (chiral centers). Examples of such diastereomers are (+)-tartaric acid and *meso*-tartaric acid. Under the present definition, compounds that would not have been included under the older definition are included: for example, *cis*- and *trans*-2-butene, *cis*- and *trans*-1,4-dimethylcyclohexane.

Diastereotopic. Used to designate apparently equivalent atoms or groups in a molecule whose replacement gives rise to diastereomers. For example: the methylene hydrogens in $CH_3CH_2CH(OH)CH_3$ are diastereotopic since the replacement of one of them by a hydroxyl group would give rise to either of the diastereomers of 2,3-dihydroxybutane. See *Enantiotopic*.

Dihedral angle (torsional angle). The angle between planes of symmetry in a molecule. More specifically, when a molecule is viewed along some bonding axis, it describes the angle between adjacent atoms or groups (not bound to the same atom). For example, the dihedral angle between the chlorine atoms in the eclipsed form of 1,2-dichloroethane is $0°$; in the *gauche* form, $\pm60°$; in the *anti* form, $\pm180°$.

Dissymmetry. See *Chirality*.

Elliptically polarized light. See under *Plane polarized light*.

Enantiomers (enantiomorphs, optical antipodes, antimers, dl-pairs). Non-superimposable mirror-image isomers.

Enantiomorphs. See *Enantiomers.*

Enantiotopic. Adjective referring to atoms or groups in a molecule that appear to be equivalent but which are not superimposable by a process of rotation. (For example, the methylene hydrogens in $CH_3CH_2CO_2H$ are enantiotopic.) The groups are so characterized if the replacement of one or the other of these groups produces an enantiomer. (For example, replacement of one of the methylene hydrogens above by bromine produces either (R)- or (S)-$CH_3CHBrCO_2H$.) The most important property of enantiotopic groups is their behavior in a chiral environment, such as in an enzymatic reaction.

Epimerization. The reversible interconversion of diastereomers, or a change in the configuration at one asymmetric carbon atom in a compound having more than one asymmetric carbon atom.

Equatorial positions. Those groups in the chair form of cyclohexane that are found directed in the average plane of the carbon atoms (that is, they are nearly perpendicular to the C_3 axis).

Equivalent atoms or groups (homotopic). When two atoms or groups in a molecule may be superimposed by the process of rotation, such that the new arrangement is indistinguishable from the old. (For example, the methylene hydrogens in $CH_3CH_2CH_3$ are equivalent.) Unlike enantiotopic groups, the replacement of one of the groups does not produce enantiomers.

Geometrical isomers. See *Cis-trans isomers.*

Hemihedral facets (dissymmetric facets, plagihedral facets). Those facets in crystals which by their presence render the crystal chiral. *Hemihedrism* may be defined as the absence of a plane, center or alternating axis of symmetry in a crystal.

Homotopic. See *Equivalent.*

Isotactic. See *Stereoregular polymer.*

Meso compounds (meso form). An achiral member of a set of stereoisomers which also contain chiral members. Its relation to the others is said to be that of a diastereomer.

Mutarotation. The spontaneous change, with time, in the optical rotation of freshly prepared solutions of certain optically active substances such that the rotation reaches an equilibrium value. Generally the mutarotation is due to an epimerization. (Example: the mutarotation of solutions of (+)- or (−)-glucose.)

Neighboring group participation (anchimeric assistance, synartetic acceleration). The participation of a group or atom near a reactive site in such a manner that it affects the stereochemistry and/or rate of the reaction.

Optical activity. The ability of a substance to refract or absorb right and left polarized light to different extents.

Optical antipodes. See *Enantiomers.*

Optical isomers. Stereoisomers that differ in their behavior towards polarized light. In recent years there has been a shift away from the use of this term because of its misleading connotations in favor of referring to compounds as stereoisomers, enantiomers, diastereomers, etc.

Optical rotation. The number of degrees by which plane polarized light is rotated by an optically active substance. Usually expressed as α_λ^t, where α is the optical rotation, t is the temperature of the substance, and λ is the wavelength of the light used. See *Specific rotation.*

Pitzer strain. See *Torsional strain.*

Plane of symmetry. See *Symmetry*.

Plane polarized light. When the oscillation of the light wave occurs in a plane perpendicular to the axis in which the light wave is traveling. Linearly (plane) polarized light can be split into two components: left- and right-handed *circularly polarized light* (each with equal phase and intensity) or *elliptically polarized light* (intensity of the two phases is unequal).

Polarimeter. An instrument used to measure the size and direction of optical rotation by optically active substances.

Racemic modification (racemic form, dl pair). A mixture containing equal numbers of enantiomers. The term 'racemic mixture' has often been used as synonymous with this term, but it is incorrect. Although the physical properties of a racemic modification in the liquid state are generally identical (except optical properties for instance) with that of the pure enantiomers, in the solid state there are deviations which depend on the phase behavior: In a *racemic compound (racemate)* the racemic modification has properties in the solid state of a distinct compound. (For example, the IR spectra may differ from that of the pure enantiomers.) In a *racemic mixture (racemic conglomerate)* the racemic modification is found to be a gross mixture of crystals of either (+) or (−) enantiomers, due to the greater affinity of an enantiomer for one of its own kind. In a *racemic solid solution* there is a random arrangement of the enantiomers. The solid has properties identical with the pure enantiomers.

Racemization. The irreversible formation of a racemic modification by the reversible interconversion of enantiomers.

Resolution. Process by which enantiomers can be separated in a racemic modification.

Rotamer. See *Conformer*.

Specific rotation (Biot's law). An equation for specifying the optical rotation at a standard concentration and cell length. $[\alpha]_\lambda^t = \alpha/ld$, where l is the path length, d is the concentration or density of the optically active substance.

Stereochemistry. The study of the relationship of the physical and chemical properties of a molecule and the three-dimensional architecture of the molecule.

Stereoisomers. Compounds having identical structures but differing arrangements of atoms in space.

Stereoregular polymers. A term used to describe the arrangement of the α-groups in the polymer formed from an α-olefin (n RCH = CH$_2$ → (RCH–CH$_2$)$_n$. When the α-groups (R) are on the same side of the polymer chain, the polymer is referred to as an *isotactic* polymer; when on alternating sides, a *syndiotactic* polymer; when randomly oriented, an *atactic* polymer.

Stereospecific reaction (stereoselective reaction). A reaction that leads to the production of only one stereoisomer. The term is usually used with reference to the way in which a reagent operates. (For example: *syn* addition, *anti* elimination, S$_N$2-inversion.)

Steric effects (steric hindrance, non-bonded interactions, Van der Waals repulsions). Those effects arising from the interpenetration of the electron clouds of non-bonding groups of atoms. When this interpenetration takes place at a reactive site so as to interfere with the functioning of a reagent, the effect is usually referred to as *steric hindrance*. The term *non-bonded interactions* is usually used in connection with the interpenetration of electron clouds that destabilizes a particular conformational isomer (for example: 1,3-non-bonded interactions in the chair of cyclohexane).

Strain theory (Baeyer strain theory, tension theory). The strain produced in cyclic compounds due to the presence of angular distortion (see *Angle strain*).

Structural isomer (constitutional isomer). Two compounds having the same constitution (i.e. same molecular formula) but differing in the sequential arrangement of atoms (disregarding the arrangement in space).

Symmetry. In modern discussions of stereochemistry, the symmetry possessed by a molecule is classified according to symmetry elements:

> *Rotational axis of symmetry, n-fold (C_n).* Where the rotation about an axis in a molecule brings the object into a position indistinguishable from its original one. The number of times this can be done in a 360° rotation equals n. (For example, ethane has two such axes, C_∞ and C_2; water has one C_2 axis; ammonia one C_3 axis.)
>
> *Plane of symmetry (σ).* Where a plane may be passed through a molecule such that half of the molecule is a reflection of the other. (Water has two such planes; ammonia, three.)
>
> *Center of symmetry.* Where there is a point within the molecule such that a straight line drawn from any part of the molecule to the center and extended an equal distance on the other side encounters an equal part.
>
> *Alternating axis of symmetry (n-fold).* When the molecule containing axis of order n is rotated by $360°/n$ about the axis and then reflection is effected across a plane at right angles to the axis, the molecule obtained is indistinguishable from the original.
>
> Any molecule possessing a plane, center, or alternating axis of symmetry is superimposable with its mirror image. Molecules having only simple axes of symmetry or molecules totally devoid of symmetry (asymmetric) will be chiral.

Synartetic acceleration. See *Neighboring group participation.*

Tension theory. See *Strain theory.*

Torsional angle. See *Dihedral angle.*

Torsional strain (eclipsing strain, Pitzer strain). The energetic barrier that arises when, for example, hydrogen atoms on adjacent carbon atoms in ethane pass by each other (as the dihedral angle approaches 0°). The barrier is thought to arise from a quantum mechanical repulsion of the bonding electrons. This barrier is not to be confused with that produced by a Van der Waals repulsion.

Van der Waals repulsion. See *Steric effects.*

Appendix C

Stereochemical satire

Kolbe's personal criticism of Van't Hoff (Chapter 5) illustrates, in a rather extreme manner, the scepticism with which many chemists of this period regarded speculations concerning the arrangement of atoms in space. While we might find Kolbe's attack amusing, undoubtedly Kolbe intended it to be taken seriously. Within a few decades, however, most of the more serious theoretical and experimental objections that had served as the basis for this scepticism had been answered, although serious criticism of stereochemical ideas continued to be voiced well into the 20th century. With the general acceptance of stereochemical theory, some chemists provided light-hearted comments in those areas where ambiguities still existed or extensions of the theory did not seem warranted.

In 1886 there appeared a short publication that had the same format as an issue of the journal *Berichte der Deutschen Chemischen Gesellschaft* (*Report of the German Chemical Society*). The special issue was entitled *Berichte der Durstigen Chemischen Gesellschaft* (*Report of the Thirsty Chemical Society*). More details about this spurious publication can be found in John Read's book *Humour and Humanism in Chemistry* and in an article by David Wilcox (see Appendix D). Read's book also provides some examples of earlier chemical satire. I am indebted to David Wilcox for providing me with an English translation (by F. R. Greenbaum and D. H. Wilcox) of the passages quoted here. The most famous article in the issue is concerned with the Kekulé 'monkey' formula of benzene, but since it is discussed elsewhere, I will not include it here. Two articles more germane to the field of stereochemistry are Communications No. 1133 and 1141:

> Communication No. 1133 by A. Speculjans: 'On the Relation of Constitution and Crystal Shape.'
> Communication No. 1141 by Wendel Schraube: 'On Polarization in Nature; Particularly in Living Organisms. Preliminary Report.'

The latter communication is too lengthy to be included here but the sense of it can be gained by noting that it is concerned with the observation of levorotatory and dextrorotatory pug-dogs:

Regarding the question of whether neutral or inactive pugs exist or may exist I intend to perform a crucial experiment in the near future. I assume that a cross breeding of pugs which have the same rotary power of levo and dextro rotary pugs should result in dogs with neutral or non-active pugs . . .

Communication No. 1133 reads as follows:

1133.   A. Speculjans: On the Relation of Constitution and Crystal Shape.

The striking differences which are found in the crystal shape of isomeric compounds have a very simple explanation when we take into consideration the various original forms of the carbon atom of which we have a certain amount of knowledge.

As is well known in organic compounds the carbon atom exists in two different forms: the tetrahedral form of Van't Hoff and the sausage form of Kekulé. Both forms may be used in many cases for the explanation of the crystal shape of organic compounds. If we assume, for example, that in an aldehyde the carbon has the sausage modification, this will explain in a most surprising way the difference between Metaldehyde and Paraldehyde.

As is well known, Metaldehyde forms long needles, while on the other hand, Paraldehyde is either a liquid or a firm porcelain-like mass in the solid state. Assuming then that the carbon atom has a sausage-like form, the aldehyde molecule must have the following shape:

Figure 1.

One can see that this form is only stable when the linkage of the two carbon atoms is sufficiently firm. However, if this linkage is not sufficiently firm, the linkage bends and through this means the molecule gets into the following shape:

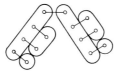

Figure 2.

When polymerization of an aldehyde occurs and the necessary firmness of the carbon compound exists, then the molecule of the polymerized compound has a still longer form. This can be easily seen if we place a number of molecules, of Fig. 1, with their ends close to each other. It explains why Metaldehyde forms so easily the long needles which are sometimes a decimeter in length.

However, in Paraldehyde we find the linkage of the carbon atom is not firm enough so that a continuous bending of the molecules occurs and this folding process increases as the chain becomes longer. This explains why Paraldehyde is a liquid at room temperature. At a lower temperature a stronger linkage does exist, but since the carbon atoms are already displaced in their positions, only a shapeless amorphous mass is formed. (Amsterdam, 7 September 1886.)

The sense of the light-hearted is obviously not lacking in the serious chemist. Some academic traditions are in fact designed to encourage such departures from the serious pursuits. Among these are the Christmas or other holiday gatherings held by graduate students. In the 19th and early 20th century the students often performed in their own plays, the nature of which are described in Read's book. Of interest to us is the playlet written by Paul Karrer (a Nobel Prize winner in chemistry in 1937) while a student in Alfred Werner's laboratory at the University of Zürich in 1911. The play, entitled: 'Drehen und Spalten' ('Rotating and Resolving'), celebrates the first resolution of an inorganic coordination compound in that year. Details about the playlet are found in an article by Kauffman (1974).

Published articles of a humorous nature relating to stereochemistry are rare in the early part of the 20th century. In 1944, however, an abstract was published in the Forensic Section of the journal *The Analyst* (Vol. 69, pp. 97–98) taken from an article entitled 'The Toxicological Significance of Levorotatory Ice Crystals; A Pharmacological Study of Acute Ice Poisoning'. The article, written by Joseph Beeman, the Director of the Oregon State Police Laboratory, originally appeared in the *Bulletin of the Bureau of Criminal Investigation, New York State Police* in December 1943 (Vol. 8, pp. 6–8). The abstract does not contain the name or initials of the abstractor (as was customary) and inquiries to date have failed to establish the identity of the abstractor or how the abstract 'slipped through' for publication. Surprisingly, the journal from which the abstract was taken does exist even though no record of it can be found in the *Chemical Abstracts*' 'List of Publications'. (The article was not abstracted by *Chemical Abstracts*.) The following is reproduced from the original article (with the permission of the Director of the New York State Police Laboratory, Captain S. Ferris):

Moussewitz[1] in his classic study of deuterium oxide in water and tissues mentions the effect of bombardment of snow crystals with the isoclonic cyclotron utilizing wave-lengths in the mega spectral region. He noticed irregularities in the extinction angles of ice crystals when the tips of the crystals were irradiated with thermal particles. Illiddennsen[2] at the same Institute noticed similar findings when the crystals were infiltrated with methyl-chloro-fluoride vapors, and was able to reduce his findings to a mathematical formula. No other reports in this field are present in the literature, and apparently the staggering significance of these findings has been unnoticed by contemporary pharmacologists.

Ice crystals are normally isotropic; that is, they are without effect upon polarized light. Under experimental conditions, the size and form of normal ice crystals can be controlled within very narrow limits, largely by strict hydrogen ion buffering. In our work, tap water was used exclusively. The water available was analyzed with a mass spectrometer to rigidly fix the concentrations of beryllium at not more than 0.0067 micrograms per liter; as is well known, concentrations of beryllium above this limit absorb therma particles, leading to irregular results. The tap water was allowed to flow into aluminum alloy trays with separators of the same metal, forming twelve 2.5 cm cubes. This particular size was chosen largely for convenience in analysis. The trays have the following composition: aluminum 65.4%, magnesium 18.7%, calcium 0.0029%, iron 5.67% and gadolinium 12.6%. The trays full of water were placed in the freezing compartment of a commercial refrigerator for six hours, the resultant ice removed manually and immediately studied on a cold stage petrographic microscope.

The crystal structure observed was a mixture of equal parts of slowly melting monoclinic rhombs and hexagonal plates. The refractive indices of the mixed crystals was 1.333. [A photograph is included in the original article which is entitled: Levorotatory Ice Crystals (100 X).] Utilizing polarized light, it was noticed that the monoclinic needle-like crystals were levorotatory, while the hexagonal plates were dextrorotatory. Separation of the two crystal types was at first done with a micromanipulator, but subsequent research indicated that the differential solubility of the materials could be effected with organic solvents. It was noted that ethyl alcohol was the most efficient solvent for the monoclinic levorotatory crystals, leaving almost untouched the dextrorotatory hexagons. Upon evaporation *in vacuo*, a 99.8% pure levorotatory ice crystal was isolated.

The levorotatory ice crystal has a needle-like shape (Fig. 1), with a point of 8 degs. on both ends. The crystal is a biaxial positive pleochroic rhomb with an extinction angle of 46 degs., a refractive index of 1.345, a melting point of −3 degs. C, a hardness of +6 and a density of 0.9996. In alcoholic solution of 10–50% strength, an alcohol–crystal complex is formed with the levorotatory compound, while the dextrorotatory compound melts innocuously.

Quantitative toxicity studies show the levorotatory compound to have a toxic index of +3.45 and the dextrorotatory compound to have a toxic index of −3.45, and that ordinary ice, when melted, consisted of a racemic mixture of the two in equal proportions, and it was apparent that the two compounds completely neutralized each other. Animals given parenteral injections of the levorotatory crystals (10 mg/kilo) developed gastritis, diarrhea, foul breath, rapid pounding pulse, bulging eyes and were extremely irritable. At autopsy, the tissues were normal upon gross examination, but microscopic studies indicated numerous sharp pointed levorotatory ice crystals sticking out of the cerebral cortex, making contact with the calvarium.

In the studies on humans, 1000 cc of a commercial alcohol complex (Brandy) was ingested in three hours in 60 cc doses with a 2.5 cm cube of ice prepared as noted above. In addition to the usual alcoholic intoxication, which in some cases was extreme, it was noted that approximately ten hours after completion of the ingestion of the drug, the subjects slept fitfully, arose with gastritis often bordering on vomiting, diarrhea, foul breath, irregular thready pulse (EKG studies indicated an inverted P–R interval), frequent odorous cructation, conjunctivitis, and sensations of heaviness in the cranial cavity; in some subjects, subjective symptoms of jabbing pains in the frontal region was noted. Nervous irritation, not relieved by thiamine, was apparent, and minor mechanical and auditory stimuli evoked atypical responses. The majority of the subjects reported that a hearty 'Good morning' or a slap on the back produced intense cranial pain. In the acute phases of poisoning, minor mechanical work of an automatic nature, such as doodling on a scratch pad, was tolerated well within the limits of normal doodling, but a definite depression of cortical function was noted. The electro-encephalographic tracing was sketchy. Memory loss was noted. Psychic functions were atypical; one subject on seeing a well-filled stocking was noted to stare glassily, look away and yawn. Minor psychoses are common, and one syndrome in particular is diagnostic: In the acutely poisoned subject, the sight and odor of an alcoholic beverage invokes reflex nausea; later, the desire for such beverages returns. A self-limiting neurosis is often noted in which the subject develops a split personality, and in this phase often has illusions that he will never again ingest alcoholic beverages; the average duration of this type of malady is from one-half to seven days. In no instance has a fatality been noted.

The course of the general drug poisoning is approximately six hours. Relief is afforded by cold milk, cold beer, acetylsalicylic acid (0.3 gm every thirty minutes), frequent naps, and a light lunch of oyster soup. Recovery is complete within twenty-four hours.

In a control group of subjects, 1000 cc of water was given in 60 cc doses with an identical amount of ice in the same period of time. No symptoms developed.

Summary:
1. Ice is a racemic mixture of toxic levorotatory and bland dextrorotatory crystals.
2. In 10–50% alcohol solution, the dextro crystals melt, while the toxic levo crystals remain.
3. The acute toxicity of the levo form is apparently due to the minute point-ed crystals protruding from the cortex to impinge upon the calvarium.
4. The acute symptoms of ice poisoning spontaneously disappear within 24 hours, apparently due to the melting of the levo form.
5. The symptoms and suggested treatment is given.

Bibliography:
1. Moussewitz, A. L., *Arch. f. Pchy. U. Norm.* 199, 276, 1933.
2. Illiddennsen, Lars, *Swenska, Norska and Finska Hellegund* 27, 645, Nov. 1939.

In 1953, D. H. R. Barton, O. Hassel, K. S. Pitzer and V. Prelog published a joint communication which sought to clarify the nomenclature in conformational analysis (Chapter 14). The proposals were widely adopted with little controversy. Some possible extensions of the terminology to cover the stereochemistry of an unusual organometallic compound were proposed in 1955 in a letter to the editor of the magazine *Chemistry and Industry* (Fig. 130). The actual author of the article was John T. Edward (now at McGill University) who recounts that he had been discussing with his colleagues over coffee the rank proliferation of new terminologies. 'I was not opposed to describing molecules as butterfly-shaped, but to introduce the term "barge" for a slightly flattened "boat" seemed to be going too far. It was only an addi-tional step to flatten the six-membered ring still further to a "raft", and all the non-sense then followed on logically. This occupied part of an afternoon when I was super-vising an undergraduate laboratory.' The article was apparently the object of a scholar-ly literary analysis in the German journal *Angewandte Chemie* (presumably in the

LETTERS TO THE EDITOR

THE STEREOCHEMISTRY OF OCTAHYDRO-HEXAIRON: A MOLECULAR " RAFT "

By Alonzo S. Smith

Chemical Laboratory, Trinity College, Dublin

SIR,—The distinction between axial (*a*) and equatorial (*e*) bonds in the " chair " form of *cyclo*hexane is now well established.[1] Lately the particular bondings in several compounds of interesting stereochemistry have been described. Dunitz and Orgel[2] have postulated that the iron atom in di*cyclo*pentadienyliron (ferrocene) is held in a " molecular sandwich." Angyal and Mills[3] have discussed the "bowsprit" (*bs*) and " flagpole " (*fp*) bonds at the " bow " and " stern " of the familiar boat form of *cyclo*hexane. Recently Beckett and Mulley,[4] besides describing a " butterfly " form for dihydroanthracene, have suggested a " barge " shape for certain molecules in which the boat shape is slightly flattened. In this configuration the *bs* bonds become *lin* (linear) and the *fp* bonds become *perp* (perpendicular).

We now wish to report that X-ray diffraction studies (to be described in detail elsewhere) on octahydrohexairon, Fe_6H_{20},[5] necessitate the introduction of another stereochemical form, the " raft " (or *r* form). In this form (I) the iron atoms are joined in a six-membered ring by the overlap of degenerate *pdf* orbitals. Six hydrogen atoms are joined to the raft by *s* (stanchion) bonds, formed by *pd* overlap, and the remaining two hydrogen atoms are held below by *k* (keelson) bonds, formed by delocalized *bhp* orbitals. On minimum this molecule assumes the conformation (II), which we designate *h* (hat), because of its resemblance to a paper hat; when this happens all *k* positions become *c* (crown) and all *s* positions become *t* (tassel).

The abnormally high density of octahydrohexairon is explained by this structure, in which a hole is left in the centre of the molecule, which accordingly would be expected to sink in water. The low density of the coordination compound, $Fe_6H_8S_2$,[6] may be explained either by a centrosymmetrical bonding of the two sulphur atoms, which serve to fill in this cavity, as in (III), or by a sideways bonding to the sulphur " buoyant balloons," as in (IV).

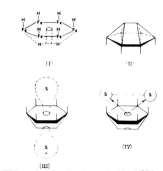

(I)   (II)

(III)

(IV)

While octahydrohexairon has no physiological properties, its unusual structure led us to test its psychological effects. Tests of the Moral Resistance Factor (m.r.f.)[7] were carried out on Wistar strain white rats lightly exposed to the compound. The rats were fed on a standard diet, and had as free choice of roughage chopped-up portions of (*a*) Beilstein's *Handbuch der Organischen Chemie* (4th edition); (*b*) the *British Medical Journal*; (*c*) *Parisian Peepshow*; and (*d*) *Supersensational Horror Comics*. Unfortunately, the compound showed a m.r.f. of 5·8, the rats having a distinct preference for (*c*) and (*d*), and after feeding going to sleep with a salacious leer.

I am, Sir, etc.,

ALONZO S. SMITH

Dublin

References

[1] Barton, Hassel, Pitzer & Prelog, *Nature, Lond.*, 1953, **172**, 1096.

[2] Dunitz & Orgel, *ibid.*, 1953, **171**, 121.

[3] Angyal and Mills, *Rev. pure appl. Chem., Aust.*, 1952, **2**, 185.

[4] Beckett & Mulley, *Chem. & Ind.*, 1955, 146.

[5] von Grizzling & Applefritter, *Proc. Bilge Acad. Sci.*, 1899, **358**, 273.

[6] Inimini and Mynimo, *J. Hotsitotsi Polytech. Inst.*, 1935, **6**, 55.

[7] Washpot & Kettleblack, *Repts., N.Y. Assoc. Alienists*, 1913, 1.

Figure 130. Reprint of a 'Letter to the Editor' in *Chemistry and Industry*, 26 March 1955, p. 354.

Nachrichten section published on 1 April) but I have been unable to locate the issues in which this analysis might have appeared.

Professor Edward indicated that numerous requests for reprints of the article were received at Trinity College. One correspondent wrote: 'Your work is undoubtedly of interest, not only to those working in the field of stereochemistry (and its related branches, such as vistavisionchemistry and cineramachemistry), but also to those who work in other fields, such as agricultural chemists . . . You mention the "molecular sandwich" of ferrocene. The combination of *cyclo*pentadiene with monomethyl astatine, or hydrogen americide, should produce even more interesting sandwiches.' Kenneth Pitzer wrote: 'I enjoyed your note very much and expect you to put a flagpole on the bowsprit of the next molecule you design which you might otherwise model on either a submarine or a helicopter.' In a more recent reconsideration of his letter he says: 'Now I would be inclined to add something about the anomaly of a raft which, when capsized, becomes a hat rather than a capsized raft.'

The Boat Form of Cyclohexane as Viewed by Midwestern Sailors

To the Editor:

The organic chemist has always felt most comfortable with a physical picture to relate to abstract ideas. This has often resulted in colorful nomenclature to describe these physical pictures. While the names for the kinds of hydrogens found on the chair form of cyclohexane were based on relationships to geometry, the whimsical vision of the "boat" form provided an outlet for chemical, classical sailors to indulge their fancy. Thus in obvious analogy to the sailing ships of yore, Angyal and Mills [*Rev. Pure Appl. Chem., Aust.*, **2**, 185 (1952)] named the hydrogens at the bow and stern of the boat the bowsprit and flagpole hydrogens, respectively.[1] Since it is not immediately evident which way the cyclohexane boat is going, the other hydrogens could equally be called the flagpole or bowsprit hydrogen.

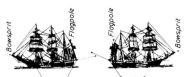

The danger of such graphic, classical nomenclature is demonstrated by this case, however, for all chemists are not good sailors. Thus the land-locked, midwestern authors, H. Hart and R. Schuetz ["Organic Chemistry: A Short Course," (4th ed.), Houghton-Mifflin, Boston, Mass., **1972**, pp. 39–40], describe an *upright* bowsprit and thus presumedly a flagpole at half-mast. The former mid-westerner Eliel ["Stereochemistry of Carbon Compounds," McGraw-Hill, New York, **1962**, pp. 205 and 207] also favors a vertical bowsprit but does fly the flag proudly in a vertical fashion. This leads to a consideration of the unfavorable van der Waals interaction denoted by this author to be the "bowsprit-flagpole" interaction [Eliel, Allinger, Angyal and Morrison, "Conformational Analysis," Interscience Publishers (Division of Wiley & Sons), New York, 1956, p. 37], which would be better classified

as the "bow and stern flagpole" interaction mentioned by Klyne [*Progress in Stereochem.*, **1**, 40 (1954)]. To ensure an equal opportunity in conformational analysis for the sailors and non-sailors, the correct designations for the hydrogens in the boat forms are given below.

In the diagram fp = flagpole; bs = bowsprit; Be = boat-equatorial; and Ba = boat-axial.

---

[1] Although A. S. Smith (*Chem. and Ind.*, 353 (1955)) credits Drs. Angyal and Mills for their introduction of these terms, Dr. Angyal in a private communication claims the use of "flagpole" only and suggests that Dr. Klyne introduced the term "bowsprit."

Gloria G. Lyle and Robert E. Lyle

University of New Hampshire
Durham, New Hampshire 03824

To the Editor:

The Lyles are certainly correct. I take full responsibility for the error in our text; your readers might like to know that Bob Schuetz is in fact a first-rate sailor, having participated for years in the annual Mackinac Straits races on Lake Michigan. W. H. Reusch of our Department tells me that nautical terms have also been used for the port and starboard hydrogens in boat cyclohexane. Be and Ba in the Lyle nomenclature being referred to as the gunwale (gunnel) and keel (*g* and *k*) hydrogens, respectively.

Harold Hart

Michigan State University
East Lansing, Michigan 48825

Figure 131. Reprint of a 'Letter to the Editor' in the *Journal of Chemical Education*, Vol. 50(9), 1973, pp. 655–656.

Professor Edward was not the only person concerned with improper terminologies used to describe the conformations of cyclohexane. The Lyles, at the University of New Hampshire, thought that the problem was geographical in origin (Fig. 131). Professor Angyal informed Dr Lyle that he believes that he and Dr Mills were the first people to use the term 'flagpole' in their 1952 review, but not the term 'bowsprit' as Alonzo S. Smith so credited them. He suggests that Klyne was the first to use the 'bowsprit' terminology. Undoubtedly this controversy merits further historical analysis.

It would appear that the readers of the journal *Chemistry and Industry* are somewhat more prone to light-hearted correspondence; or perhaps the editors are more tolerant toward the publication of such letters. The following letter was published in the 25 August 1962 (p. 1533) issue of *Chemistry and Industry*:

SEPARATION OF ENANTIOMORPHS
BY GAS CHROMATOGRAPHY

Sir, — The gas chromatographic separation of enantiomorphs is the object of intensive investigation, as revealed by several recent communications in *Chemistry and Industry*. Accordingly, we think it appropriate to make known our approach to this problem.

Use of optically active column material is the essential feature of previous work. It occurred to us some time ago that a more effective separation might be achieved if this principle could be combined with the use of mirrors, the latter having been associated with stereochemistry for many years. We were led to this view by a consideration of the mechanism of adsorption at a mirrored surface. The vacant sites adjoining an adsorbed molecule of the *d*-form would appear to be occupied by molecules of the *l*-form. Fewer adsorption sites would thus be available to the *l*-form, which as a consequence should be eluted first from the column. Evidently, a modification of this hypothesis would account also for the prior emergence of the *d*-form.

We consider that a satisfactory column for work of this nature might be packed with glass micro-beads, mirrored by the thermal decomposition of dextrorotatory methyl sec.-butylmercury. An elaboration of the theory will soon appear[1] and it is hoped that further details will be published when experimental confirmation has been obtained.

Yours faithfully,
WILLIAM GRAHAM

41, Sturgess Road,
    Reading,
        Massachusetts, USA

Reference
  1. *J. Cuchinnation*, in press.

My research to date has revealed only a few publications of the sort that appeared in *Chemistry and Industry* and *The Analyst*. The 1 April issue of the *Nachrichten aus Chemie und Technik* section of *Angewandte Chemie* usually contains articles of possible interest in this regard, however. In recent years articles have appeared based on addresses given at the Bürgenstock Stereochemistry Conferences. The reader is particularly encouraged to consider the article by Professor A. Troischose [André Dreiding, University of Zürich] entitled 'Methin und seine Sphäromerisierungen' since some interesting proposals are made as regards the synthesis of tetrahedrane and dodecahedrane (Chapter 16). The article appears in the 1970 issue (Vol. 18 (7), pp. 127–128).

Professor Derek Davenport (from Purdue University) has provided the only known example of a stereochemical 'spoonerism'. As background, it should be pointed out that Samuel Smiles should rightfully be given credit for the first synthesis of an optically active sulfonium compound in 1900 (Chapter 10). Smiles' report (*J. Chem. Soc.*, 1174–1179, 1900) appeared after the report of the synthesis of a similar compound by Wm. J. Pope and S. J. Peachey (*J. Chem. Soc.*, 107–175, 1900). The synthesis of the later compound apparently took place after the results of Smiles' work had been communicated to Pope. Thereafter, whenever Smiles would speak about the preparation of the optically active sulfonium compounds, he would state: 'These were discovered by Peep and Poachy – I beg your pardon – by Pope and Peachey.' (Smiles was reported to be a very articulate and precise speaker.)

The reader (or reviewer) of this appendix may well ask whether its inclusion in the book contributes to our understanding of the development of stereochemistry. I can only answer that it probably does, but no attempt has been made to support this belief. The appendix was included for the enjoyment of the reader and will not be subjected to any rigorous historical analysis at this time. There is always that possibility for the future as more research sources become available.

"*Parkinson! Do you mind?*"

Figure 132. (From *Punch*, 16 September 1959, p. 69.)

Appendix D

Supplementary reading

I. *Stereochemistry and conformational analysis*

1. K. Blaha, O. Cervinka and J. Kovar, *Fundamentals of Stereochemistry and Conformational Analysis*. Iliffe Books, London (1971).
2. J. Dale, *Stereochemistry and Conformational Analysis*. Verlag Chemie, New York (1978).
3. E. L. Eliel, N. L. Allinger, S. J. Angyal and G. A. Morrison, *Conformational Analysis*. Wiley-Interscience, New York (1965).
4. E. L. Eliel, *Stereochemistry of Carbon Compounds*. McGraw-Hill, New York and London (1965).
5. E. L. Eliel, *Elements of Stereochemistry*. Wiley, New York (1969).
6. R. W. Geise, R. P. Mikulak and O. A. Runquist, *Stereochemistry: An Introductory Programmed Text*. Burgess, Minneapolis (1976).
7. F. D. Gunstone, *Basic Stereochemistry*. English Universities Press, London (1974).
8. F. D. Gunstone, *Guidebook to Stereochemistry*. Longman, Inc., New York (1975).
9. G. Hallas, *Organic Stereochemistry*. McGraw-Hill, New York and London (1965).
10. M. Hanack, *Conformation Theory*. Academic Press, New York (1965).
11. H. Kagan, *Organic Stereochemistry*. Edward Arnold, London (1979).
12. K. Mislow, *Introduction to Stereochemistry*. Addison-Wesley/Benjamin, New York (1965).
13. H. B. Mosher, *Organic Stereochemistry*. ACS Audio Course, American Chemical Society, Washington, DC (1980).
14. G. Natta and M. Farina, *Stereochemistry*. Longman, London; Harper and Row, New York (1972).
15. B. Testa, *Principles of Organic Stereochemistry* (Vol. 6 in Studies in Organic Chemistry Series). Marcel Dekker, New York (1978).

II. *Serial publications*

1. B. J. Aylett and N. M. Harris, *Progress in Stereochemistry*, Vols. 1–4. Butterworths/Plenum, London (1954–69).
2. E. L. Eliel and N. L. Allinger, *Topics in Stereochemistry*, Vols. 1–11. Wiley-Interscience, New York (1967–79).
3. F. Boschke (ed.), *Topics in Current Chemistry*. Springer-Verlag, New York:
    Vol. 15, *Dynamic Stereochemistry* (1970).
    Vol. 31, *Stereo- and Theoretical Chemistry* (1972).
    Vol. 47, *Stereochemistry I* (1974).
    Vol. 48, *Stereochemistry II* (1974).
    Vol. 49, *Symmetry and Chirality* (1974).

III. *Miscellaneous books on stereochemistry and its applications*

1. W. L. Alworth, *Stereochemistry and Its Applications to Biochemistry: The Relation Between Substrate Symmetry and Biological Stereospecificity*. Wiley-Interscience, New York (1972).
2. R. Bentley, *Molecular Asymmetry in Biology*, Vols. 1–2. Academic Press, New York (1969–70).
3. W. L. F. Armarego, *Stereochemistry of Heterocyclic Compounds, Parts 1 and 2*. Halsted Press, New York (1977).
4. I. L. Finar, *Organic Chemistry*, Vol. 2: *Stereochemistry and the Chemistry of Natural Products*, 5th edn. Longman, Inc., New York (1975).
5. J. F. Stoddart, *Stereochemistry of Carbohydrates*. Wiley-Interscience, New York (1971).
6. J. D. Morrison and H. S. Mosher, *Asymmetric Organic Reactions*. American Chemical Society, Washington, DC (1976).
7. M. S. Newman (ed.), *Steric Effects in Organic Chemistry*. Wiley, New York (1956).
8. D. Whittaker, *Stereochemistry and Mechanism*. Clarendon Press, Oxford (1973).
9. I. Bernal, W. C. Hamilton and J. S. Ricci, *Symmetry: A Stereoscopic Guide for Chemists*. Freeman, San Francisco (1972).
10. J. D. Donaldson and S. D. Ross, *Symmetry and Stereochemistry*. Halsted Press, New York (1972).
11. R. J. Gillespie, *Molecular Geometry*. Van Nostrand Reinhold, New York (1972).
12. C. C. Price, *Geometry of Molecules*. McGraw-Hill, New York (1971).
13. A. Walton, *Molecular and Crystal Structure Models*. Halsted Press, New York (1978).
14. H. B. Kagan (ed.), *Determination of Configurations by Spectrometric Methods*. Vols. 1–3. Heyden, London (1977).
15. H. B. Kagan, *Absolute Configurations of 6000 Selected Compounds with One Asymmetric Carbon Atom*, Vol. 4. Heyden, London (1977).
16. W. Klyne and J. Buckingham, *Atlas of Stereochemistry: Absolute Configurations of Organic Molecules*, 2nd edn (1st edn, 1974). Oxford University Press, Oxford (1978).

IV. *Selected publications that provide historical background to the development of stereochemistry and conformational analysis* (Many of the publications in I–III also contain valuable contributions to the history of stereochemistry.)

1. O. T. Benfey, *Classics in the Theory of Chemical Combination*. Dover, New York (1963). (This book contains reprints of important papers of a number of 19th-century chemists: F. Wöhler and J. Liebig, A. Laurent, A. W. Williamson, E. Frankland, A. Kekulé, A. S. Couper, J. H. van't Hoff and J. A. Le Bel.)
2. O. T. Benfey, *From Vital Force to Structural Formulas*. Houghton Mifflin, Boston; American Chemical Society, Washington, DC (1964).
3. O. T. Benfey (ed.), *Kekulé Centennial* (Advances in Chemistry Series, No. 61). American Chemical Society, Washington, DC (1966). (Contains 10 articles concerned with the historical development of structural chemistry and stereochemisty.)
4. J. D. Bernal, *Science and Industry in the Nineteenth Century* (Chapter VII, 'Molecular Asymmetry: Antecedents and Consequences of Pasteur's Discovery of Molecular Asymmetry'). Indiana University Press, Bloomington (1953).
5. A. J. Ihde, *The Development of Modern Chemistry*. Harper & Row, New York (1964).
6. H. M. Leicester and H. S. Klickstein, *A Source Book of Chemistry, 1400–1900*. McGraw-Hill, New York (1952). (Contains excerpts from key papers of some 83 chemists.)
7. S. H. Mauskopf, *Crystals and Compounds: Molecular Structure and Composition in Nineteenth-Century French Science* (Vol. 66 (n.s.), part 3, in the 'Transactions of the American Philosophical Society'). American Philosophical Society, Philadelphia (1976).
8. W. G. Palmer, *A History of the Concept of Valency to 1930*. Cambridge University Press, Cambridge (1965).
9. J. R. Partington, *A History of Chemistry*, Vol. 4. Macmillan, London and New York (1964).
10. O. B. Ramsay (ed.), *Van't Hoff–Le Bel Centennial* (ACS Symposium Series, No. 12). American Chemical Society, Washington DC (1975). (Contains 13 articles concerned with the history of stereochemistry.)
11. G. M. Richardson, *Foundations of Stereochemistry*. American Book Company, New York (1901).
12. C. A. Russell, *The History of Valency*. Humanities Press, New York; Leicester University Press, Leicester (1971).
13. H. Gilman (ed.), *Organic Chemistry*. Wiley, New York (1943). (The chapter by R. L. Shriner and R. Adams on 'Optical Isomerism' provides an excellent review of the history of stereochemistry from the perspective of these two well-known chemists. The first edition may also be usefully consulted.)
14. It is not possible to include a list of relevant journal articles. The reader is urged to consult the following journals in particular: *The Journal of Chemical Education*, *Chemistry*, and *Isis*. In 1974 a number of chemistry journals published special Van't Hoff–Le Bel Centennial articles or issues: *Tetrahedron*, Vol. 30 (No. 12–13); *Chemiker Zeitung*, Vol. 97; *Angewandte Chemie, International Edition*, Vol. 13; *Chemie im unserer Zeit*, Vol. 8; *Chemical Technology*, 2 December.

V.   *Some background sources relating to 'chemical' and 'stereochemical' humour and satire*

1.   R. E. Oesper, *The Human Side of Scientists.* University of Cincinnati, Cincinnati (1975).
2.   J. Read, *Humour and Humanism in Chemistry.* G. Bell, London (1947).
3.   R. L. Weber, *A Random Walk in Science.* Institute of Physics, London (1973).
4.   The following articles may be of interest:
     (a) G.G. Kauffman and H. K. Doswald, 'Rotating and Resolving: A Tragicomic Popular Play' (written by Paul Karrer). *Chemistry*, Vol. 47, pp. 8–12 (1974).
     (b) D.H. Wilcox, 'Kekulé's Benzene Ring Theory – A Subject for Light-Hearted Banter'. *J. Chemical Education*, Vol. 42, pp. 266–267 (1965).

Subject Index

Abietic acid, 188
Absolute asymmetric synthesis, 198
Absolute configuration, 231
Acetylene, Kekulé-Baeyer model, 65
Achiral, 117, 231
Active site, 219
Acyloin synthesis, 25
Adamantane, 5, 168
1-Adamantyl carbenium ion, 213
Addition reactions, 98ff, 106
Alkene structures, 86, 89
Allenes (chiral), 118, 127
Alloisomerism, 101
Ammonia inversion, 132
Amylene isomers (Le Bel), 89
Anchimeric assistance, 145, 231
Angle strain, 191, 231
Ansa compounds, 157
Apocamphyl chloride, 114, 144
Asparagine, 73
Aspartic acid, 74
Asymmetric carbon atom, 83, 91, 117, 231
Asymmetric destruction, 194
Asymmetric induction, 196
Asymmetric methyl, 38
Asymmetric reactions, 119
Asymmetric synthesis 194, 231
Asymmetry, 77, 116
Atactic polymers, 225, 231
Atomism, 45–47, 93
Atropoisomerism, 153, 157, 231

Axial positions, 184, 231
Axis of symmetry, 231

Baeyer strain, 2, 191, 231
Barrelene, 205
Basketene, 205
Benzaldoxime, 131
Benzene, formulas, 57, 60, 62, 64, 164, 206
β-Benzenehexachloride, 169, 183, 187
Benzidine, derivatives, 154
Benzil, dioxime, 129
Benzvalene, 206
Betweenanone, 204
Bicyclobutonium ion, 145
Bicyclohexane isomers, 169
Bimolecular eliminations (E_2), 182
Bimolecular nucleophilic substitution (S_N2), 113, 153
Binaphthyl (chiral), 200
Biot's law, 231
Biphenyl, 200
Bond, chemical, 50
Boric acid–polyol complexes, 173
Bredt's rule, 5, 34, 204
2-Bromopropanoic acid, 109, 111
Bullvalene, 208

Camphene hydrochloride, 144
Camphor, 168
Carbanions, 146
Carbocations, 143

Carbenium ion, 231
Carbonium ions, 142, 144, 213
Carboxypeptidase, 220
Carvone, 214
Catenanes, 208
Chemical structure, 55, 56
Chiral axis, 127
Chiral centre, 231
Chiral methyl, 224
Chirality, 77, 116, 121, 232
Chloroamphenicol, 215
1-Chloroapocamphene, 114
Chlorocyclohexane, 166
Chloroiodomethanesulfuric acid, 91
Chlorosuccinic acid, 108
Cholesterol, 188
Cholic acid, 186
Chymotrypsin, 219
Citric acid cycle, 215
Combining power, 53
Configuration, absolute, 128, 232
Configuration, CIP, 119, 126
Configuration, D/L, E/Z, R/S, 123–127
Configuration, inversion/retention of,
    108
Conformation, 1, 85, 149, 165, 187,
    192, 232
Conformational analysis, 1, 4, 173, 188,
    232
Conformers, 212, 232
Congressane, 205
Constellation, 6, 30
Constitutional formula, 232
Corpuscular philosophy, 46
Cram's rule, 197
Crown ethers, 222
Crystallography, 47, 51, 77
Cubane, 205, 206
Cuneane, 205
Cycloalkane stabilities, 166
Cycloalkanediol conformations, 174
Cyclobutane synthesis, 160
Cyclodextrins, 221
Cyclohexaamylose, 221
Cyclohexane, 162, 169, 175
Cyclohexane-1, 2-dicarboxylic acid, 164
Cyclononanone, attempted resoln, 170

Cyclopentane, 31, 160
Cyclophanes, 203
Cyclopropane, 160

Decalin, 168
Dewar benzene, 64, 206
Diastereotopic, 218, 232
2, 3-Dibromobutane, 105
Dibromoethane isomers, 66
1, 3-dibromopropane, 157
Dibromosuccinic acid, 101
1, 2-Dichloroethane, 62, 180
Dimethylalkyl pyrophosphate, 224
Dimethylmaleic acid dianion, 104
1, 2-Diphenylethane-1, 2-diol, 178
Dipole moment, 3, 179, 183
Dissymmetry, 77, 117, 232
Dodecahedrane, 206
Dutch liquid, 62

Electron diffraction, 3, 183
Elements, 44ff
Elimination, bimolecular (E$_2$), 20, 100,
    105
Enantiomers, 73, 83, 95, 233
Enantiomorphs, 73, 233
Enantiotopic, 218
Equatorial positions, 3, 16, 184, 233
Esterification, law of, 151
Ethylene bromination, 103
(Et)(i-Bu)(Me)(Pr)N^{1+} Cl^{1-}, 135
(Et)(Me)(Pr)N, 130

Fluxional molecules, 208
Formulas
    of benzene, 58
    of Brown, 60
    configurational, 122
    of Dalton, 47
    of Fischer, 123
    glyptic, 63
    graphic, 57
    Kekulé, 57ff
    Type, 53, 80
    of Wollaston, 46
Free radicals, 145
Fumaric acid, 86, 90, 98

Geometrical isomers, 80, 86
Glucose configuration, 124
Glyceraldehyde configuration, 123
Glyptic formulas, 62
Graphic formulas, 57
Gulose configuration, 124

Handedness, 116
Hantzch-Werner hypothesis, 119, 130, 148
Heats of formation, 166
Hexahelicene, 157
Hexahydrohomophthalic acid, 168
Hexahydrophthalic acid isomers, 164ff
Homocubane, 205
Hooplanes, 209
Hydrobenzoin-acetone ketal, 178
Hypostrophene, 205

Ice crystals (levorotatory), 238
Inorganic complexes (chiral), 141
Inverse square law (Biot), 69
Inversion of configuration, 112
Inter-, intra-radial effects, 186
Iodooctane, S_N2 reaction, 113
Isobutylene, 48
Isodibromosuccinic acid, 101
Isopentenyl pyrophosphate, 8, 224
Isomerism, 48, 80, 101
Isomorphism, law of, 71
Isotactic polymers, 222, 233

'Key and lock' theory, 196, 215
Kinetic method of resolution, 196
α-Ketoglutaric acid, 215
Krebs' cycle, 215

Lactic acid, 78, 126
'Lock and key' theory, 196, 215

Macrocyclic compounds, 172
Malate synthase, 38
Maleic acid, 86, 90, 98, 102
Malic acid, 74, 107, 225
Mandelic acid, 196
Menthol stereoisomers, 185, 189, 196
Methyl, chiral, 224

Mevalonic acid, 223
Meyer-Auwers theory, 130
Models, atomic-molecular
  croquet-ball, 62
  Dalton, 49
  Derx, 63
  Dewar, 63
  glyptic, 62
  Kekulé, 64, 161, 168
  Koerner, 65
  Pasteur, 73
  spring-bond, 163
  Stuart, 65, 170, 185
  Van't Hoff, 86
Molecular 'rafts', 241
Muskone, 172

Nitrogen oxides, 137
Nucleophilic substitution (S_N1,S_N2), 113

Octahydrohexairon, 241
Ogston effect, 217ff
Olefiant gas, 241
Optical activity, 69ff
Oxaloacetic acid, 216

Paracyclophanes, 157, 203
Paralactic acid, 78
Paratartaric acid, 71
Pentavalent nitrogen, 133ff
Pitzer strain, 182, 191, 233
Platonic molecules, 206
Polar positions, 16
Polarized light, 69
Prelog's rule, 197
Prismane, 206
Propellane, 207
Pug-dogs, chiral, 236
Pyruvic acid, 215

Quadricyclene, 205
Quartz, 69ff
Quaternary ammonium salts, 133, 135

Racemic acid, 71, 76
Radical theory, 52
Radicals, free, 145ff

Rafts, molecular, 241
Raman spectroscopy, 10, 180, 183
Resolution, 75, 194
Rotation, free, 86
Rotational barrier, 31, 148, 156, 179

Saccharic acid configuration, 123
Sarcolactic acid, 78
Spiranes (chiral), 120
Spoonerism, stereochemical, 244
Squalene, 223
Stereochemistry, 8, 43, 98, 224
Stereochemistry, hidden, 8, 224
Steric hindrance, 20, 150
Stilbene bromination, 103
Stilbene dichlorides, 179
Strain, 'B' and 'F', 152
Strain theory (Baeyer), 2, 95, 161, 235
Strontium formate, chiral, 74
Stuart models, 170, 185
Sulfinic acid esters, chiral, 141
Sulfonium cmpds., chiral, 140
Sulfoxides, chiral, 141
Symmetry, 116ff, 235
Synartetic assistance, 145, 235
Syndiotactic polymers, 225

Tartar, 71
Tartaric acid, 71, 76, 89, 179, 194
Tension theory (Baeyer), 235
Tetrahedral carbon
   Butlerov, 67
   irregular, 95
   Kekule, 64ff
   Koerner, 65

Tetrahedral carbon – cont.
   Van't Hoff, 83ff
   Victor Meyer, 68, 83
Tetrahedrane, 205, 206
Tetraatomicity, carbon, 54
Tetravalency, carbon, 54, 67
   planar config., 210
Three-point receptor site, 217
Torsional strain, 182, 191, 235
Triansa compounds, 209
Triasylcarbonium ions, 212
Tricarboxylic acid cycle, 215
Triphenylmethane stereoisomers, 211
Trivalent nitrogen, 110
Troeger's base, 5, 132
Type formulas, 53, 80
Twistane, 205

Urea isomerism, 48

Valence shell election repulsion, 97
Valency-deflection hypothesis, 171
Valency theory
   Couper, 54ff
   Kekule, 54ff
   Werner, 94, 109
Van der Waals forces, 191

Wagner-Meerwein rearrangement, 144
Walden inversion, 102, 105ff, 112
   inorganic, 114

X-ray crystallography, 96, 168, 169,
   183
X-ray diffraction, 3

Name Index

Adams, R., 144, 145, 156, 167
Abrecht, H., 146
Alexander, E.R., 121
Allinger, N.L., 191
Alworth, W., 218
Ampere, G.M., 51
Anschutz, R., 100
Arigoni, D., 9
Arnett, E., 153
Aschan, O., 118, 167
Aston, J.G., 184
Avogadro, A., 50
Auwers, K.F. von, 94, 129

Backer, H.J., 120
Baeyer, 9, 65, 91, 122, 160, 169, 172
Bailar, J., 114
Bain, A.M., 119, 131
Bak, B., 183
Bartlett, P., 103, 114, 144, 146
Barton, D.H.R., 1, 3, 16, 175, 184, 187, 240
Bastiansen, O., 187
Beckmann, E.O., 129
Beeman, J., 238
Bell, F., 156
Bender, M., 221
Berk, J., 175
Berthelot, M., 84
Berzelius, J.J., 48
Bulman, E., 110, 143
Bijvoet, J.M., 128

Bilicke, C., 169, 183
Birch, A.J., 189
Bischoff, C.A., 134, 148
Biot, J.B., 69, 81
Blow, D.M., 220
Boekelheide, V., 203
Boeseken, J., 173
Botlini, A.T., 132
Boyle, R., 45
Bragg, W.H. & W.L., 45
Bredt, J., 204
Brois, S.J., 132
Brown, A.C., 60, 61, 91
Brown, H.C., 152
Buckel, W., 41
Bushweller, C.H., 166
Butlerov, A.M., 55, 56, 67

Cahn, R.S., 116
Cain, J., 194
Campbell, A.N., 96
Cannizzaro, S., 52, 68
Clark, J., 179
Cloton, N., 206
Cole, T.W.Jr., 206
Cornforth, J.W., 5, 7, 38, 233
Coulthard, A., 154
Couper, A.S., 54, 60, 61
Cram, D.J., 144, 147, 197, 220
Curie, P., 198
Cushing, A., 215

Dalton, J., 46, 47, 49
Davis, T., 200
Derx, H.G., 174
Descartes, R., 45
Dessaignes, M., 75
Dewar, J., 63, 64
Doering, W. von E., 147
Donath, W.E., 184
Dostrovsky, I., 187
Dumas, J.B.A., 53

Easson, L., 215
Eaton, P., 206
Edward, J.T., 240
Eggerer, H., 41, 225
Eichelberger, L., 102
Eiloort, A., 91, 178
Eliel, E.L., 121, 190
Epicurus, 44
Eyring, H., 180, 183

Faraday, M., 48
Fieser, L., 5, 186
Fischer, E., 95, 109, 160, 195, 215
Fittig, G.C., 59
Francis, A.W., 103
Frankland, E., 53, 56
Frankland, P.F., 105, 110
Freiberg, L.A., 191
Freudenberg, K., 124

Gassendi, P., 45
Gaudin, M.-A., 50
Gay-Lussac, J.L., 71
Gerhardt, C.F., 77
Gillespie, R.J., 97
Gilman, H., 167
Glass, M.A., 170
Goldschmidt, H., 129
Gomberg, M., 142, 144
Graham, W., 243
Gutschow, C., 41
Guyé, P.A., 91

Hammett, L.P., 152
Hantzch, A., 130
Harrison, I.T. and S., 208

Hart, H., 242
Hassel, O., 1, 10, 183, 187, 240
Haüy, R.J., 47, 70
Havinga, E., 199
Havrez, P., 61, 62
Haworth, H.N., 181
Hazebroek, P., 192
Hemingway, A., 215
Henderson, R.M., 220
Hendricks, S.B., 169
Hendrickson, J., 193
Henry, L., 95
Hermans, P.H., 149, 174, 177
Herschel, J.F.W., 70
Hofman, R., 210
Hoffmann, A., 62, 133, 150
Holness, N., 178, 190
Hückel, W., 168, 191
Hughes, E., 105, 111, 113, 143, 153,
    187
Huygens, C., 47, 69

Ingold, C.K., 105, 111, 113, 116, 143,
    153, 171, 187

Jaeger, F.M., 117
Janson, S.E., 120
Japp, F.R., 198
Jaray, A.S., 201
Jensen, F.R., 166
Jones, H.O., 136
Jungfleisch, E., 195

Kaiser, E.T., 221
Kampf, A., 155
Karagounis, G., 146
Katz, T.J., 206
Käufler, F., 153
Kehrmann, F., 150
Kekulé, F.A., 54, 64, 79, 93, 99, 133,
    161, 168, 236
Kellas, A., 151
Kelvin, Lord, 117
Kemp, J.D., 180
Kenner, J., 155
Kenyon, J., 156
Kestner, C., 71,
Kilpatrick, J., 184

Kimball, G., 104
Kindler, K., 153
King, H., 155
Kipping, F.S., 131, 199
Kistiakowsky, G.B., 180
Kleber, C., 166
Klyne, W., 126, 182
Knox, L.H., 114, 144
Koerner, W., 65
Kohler, E.P., 120
Kohlrausch, K.W.F., 183
Kolbe, A.W.H., 78, 91, 236
Kornblum, N., 146
Krebs, H., 216
Kuhn, R., 146
Kuhn, W., 200

Langmuir, I., 140
Langseth, A., 10
Laurent, A., 51
Le Bel, J.A., 43, 87, 112, 130, 134, 195
Lecco, M., 133
Lednicer, D., 157
Le Fevre, R.J.W., 155
Leucippus, 44
Lewis, G.N., 97, 112, 138
Lieben, A., 68
Liebig, J., 53
Linnett, J., 97
Longseth, A., 183
Loschmidt, J., 61
Lucretius, 44
Lukach, C.A., 178, 190
Lüttringhaus, A., 209
Lyle, G.C. & R.E., 242

Magnanini, 173
Maier, G., 206
Maitland, P., 119
Malus, E.L., 69
Marckwald, W., 135, 196
Mark, H., 96
Markovnikov, V., 160
Marsh, J., 119
Marshall, J., 204
Marvel, C., 138, 167, 170
Mazur, R.H., 145
McCasland, J.E., 118

McKenzie, A., 102, 196
Meer, N., 113
Meisenheimer, J., 110, 132, 137
Meulenhoff, J., 175
Meyer, J.L., 52
Meyer, V., 43, 68, 94, 129, 133, 148, 150, 157, 165
Michael, A., 101
Micklethwait, F., 154
Mills, W.H., 119, 131, 138, 156
Mislow, K., 210
Mitscherlich, E., 71
Mizushima, S., 180
Mohr, E.W.M., 2, 118, 167, 187, 191

Naquet, A., 61
Nevill, T.P., 144
Newman, M., 149, 153, 157, 181
Newton, I., 46
Nier, A.O., 215
Nyholm, R., 97

Ogston, A.G., 217
Olah, G., 213
Olson, A.R., 114
Oosterhoff, L.J., 192, 213

Paquette, L.A., 206
Pasteur, L., 72, 194, 198
Paterno, E., 66
Pauling, L., 96, 114, 183
Peachey, S.J., 135, 140, 244
Perkin, W.H., 119, 160, 167
Pfeiffer, P., 110
Pfriem, S., 206
Phillips, H., 111, 141
Pickering, S.V., 136
Pickett, L.W., 179
Pinkus, A.G., 121
Pitzer, K., 31, 175, 180, 184, 240
Polyanyi, M., 114, 153
Pope, W.J., 91, 119, 120, 135, 140, 199, 244
Popjak, J., 8, 224,
Prager, W.L., 151
Prelog, V., 5, 25, 116, 132, 182, 184, 189, 197, 204, 240

Pressman, D., 105
Proskow, S., 118
Provastaye, F.H. de la, 71

Read, J., 91, 135, 185, 189, 236
Redmond, J.W., 41
Reeves, R., 189
Retey, J., 9
Reverez, C., 168
Riecke, E., 148
Roberts, I., 104
Roberts, J.D., 132, 144, 145
Robinson, R., 7
Rosanoff, M.S., 124, 151
Ruzicka, L., 5, 25, 172

Sachse, H., 2, 165, 174, 191
Sales, E. de, 144
Salway, A.H., 131
Scheele, C.W., 71, 78
Schill, G., 209
Schlenk, W., 138
Schmidt, J., 155
Schraube, W., 236
Schurink, H.B.J., 120
Senter, G., 111
Shriner, R., 167
Sidgwick, N.V., 140
Slater, R.H., 114
Smiles, S., 140, 244
Smith, A.S., 241
Speculjans, G., 236
Spitzer, R., 184
Stedman, E., 215
Steitz, T.G., 220
Stohman, F., 166
Stoll, M., 6, 172
Stoll-Compte, G., 172
Stuart, H.A., 170
Stubbing, W.F., 155
Sudborough, J., 151

Toft, R., 153
Tarbell, S., 103

Terry, E., 102
Thomson, T., 49
Thorpe, J.F., 171
Tischler, M., 120
Twiner, E.E., 155

Van Loon, C., 174
Van't Hoff, J.H., 43, 82, 118, 148, 173, 200
Vaubel, W., 94

Walden, P., 107
Walden, R., 210
Walker, J.T., 120
Wallach, O., 119
Wallis, E.S., 144, 146
Weng, E.J., 206
Warren, E.H., 138
Wasserman, E., 208
Watson, H.B., 152
Weissberger, A., 179
Weissenberger, K., 96
Werkman, C.H., 215
Werner, A., 94, 109, 130, 136, 141, 167
Westheimer, F., 156, 187
Wheland, G., 117
Whitmore, F., 153
Whyte, L.L., 117
Wieland, D., 132
Wightman, W.A., 169, 191
Wilbrand, J., 59
Wilcox, C.F. Jr., 210
Wilson, C.L., 144
Windaus, A., 168
Winstein, S. 105, 144, 178, 190
Wislicenus, J.G., 78, 91, 98, 165
Wohl, A., 125
Wöhler, F., 48
Wolf, K.L., 179
Wollaston, W.H., 46
Wood, H.G., 215
Woodward, R.B., 4
Wurtz, C.A., 59, 78, 81